BEADING WITH CRYSTALS

36 Simply Inspired Jewelry Designs

LARK JEWELRY
& BEADING

BEADING *with* CRYSTALS

A LARK BESTSELLER
NOW IN PAPERBACK!

36 SIMPLY INSPIRED JEWELRY DESIGNS

Jean Campbell and Katherine Aimone

BEADING WITH CRYSTALS

36 Simply Inspired Jewelry Designs

Katherine Duncan Aimone & Jean Campbell

LARK CRAFTS
Asheville

Development & Coordinating Editor
Katherine Duncan Aimone

Technical Writer & Editor
Jean Campbell

Art Director
Stacey Budge

Cover Designer
Carol Morse Barnao

Associate Art Directors
Travis Medford
Lance Wille

Art Production Assistant
Jeff Hamilton

Editorial Assistance
Delores Gosnell

Illustrators
J'aime Allene
Bonnie Brooks

Photographer
Stewart O'Shields

LARK CRAFTS

An Imprint of Sterling Publishing
387 Park Avenue South
New York, NY 10016

If you have questions or comments about
this book, please visit: larkcrafts.com

The Library of Congress has cataloged the hardcover edition as follows:

Duncan-Aimone, Katherine
  Beading with crystals : beautiful jewelry, simple techniques / Katherine
Duncan Aimone & Jean Campbell.
    p. cm.
  ISBN-13: 978-1-60059-036-8 (hc-plc with jacket : alk. paper)
  ISBN-10: 1-60059-036-5 (hc-plc with jacket : alk. paper)
  1. Beadwork--Patterns. 2. Jewelry making. I. Campbell, Jean, 1964- II.
Title.
  TT860.D845 2007
  745.594'2--dc22

                                        2006101671

10 9 8 7 6 5 4 3 2 1

Published by Lark Crafts, An Imprint of
Sterling Publishing Co., Inc.
387 Park Avenue South, New York, N.Y. 10016

First Paperback Edition 2012

© 2007, Lark Crafts, an Imprint of Sterling Publishing Co., Inc.

Distributed in Canada by Sterling Publishing,
c/o Canadian Manda Group, 165 Dufferin Street
Toronto, Ontario, Canada M6K 3H6

Distributed in the United Kingdom by GMC Distribution Services,
Castle Place, 166 High Street, Lewes, East Sussex, England BN7 1XU

Distributed in Australia by Capricorn Link (Australia) Pty Ltd.,
P.O. Box 704, Windsor, NSW 2756 Australia

Manufactured in Canada

ISBN 13: 978-1-60059-036-8 (hardcover) 978-1-4547-0360-0 (paperback)

For information about custom editions, special sales, and premium and corporate
purchases, please contact Sterling Special Sales Department at 800-805-5489 or
specialsales@sterlingpub.com.

Requests for information about desk and examination copies available to college and
university professors must be submitted to academic@larkbooks.com. Our complete policy
can be found at www.larkcrafts.com.

Table of Contents

When Marilyn Monroe donned a Swarovski-laden gown on the famous occasion of John F. Kennedy's 1962 birthday, she undoubtedly recognized the power of crystals. Let's face it…there's not much that can rival 6,000 shimmering crystals on a dress!

Crystal beads continue to be a huge part of our fashion world, whether valued for their "bling" factor, their boney elegance and inherent glamour, or their unquestionable ability to turn heads. This inherent appeal, coupled with the already enormous popularity of beading, has resulted in a kind of crystal frenzy!

The 25 jewelry designers featured in this book all relish working with crystals. In fact, most of them admit they're hopelessly hooked. As you peruse the 36 designs in this book, you'll find yourself equally inspired by the alluring qualities of these brilliant beads.

Necklaces, bracelets, cuffs, brooches, and earrings sparkle with the creativity of their makers. Preview Tamara Honaman's fringed bracelet (page 91), punctuated with smoky topaz and gold crystal beads and finished with a showy crystal clasp. Or catch a glimpse of Wendy Witchner's "bling" rings (page 48), which are playful, stylish, and yet versatile enough to be worn with anything from a gown to jeans. Feast your eyes on Candie Cooper's gorgeous necklace of beaded beads (page 135), which elevates crystals to a new dimension. Be convinced that crystals are like no other material when you look at Elizabeth Larsen's amazing wirework and crystal bracelet (page 117). And prepare to be dazzled by Katherine Song's crystal and gold necklace (page 103), which underscores the classic splendor of clear crystals.

These projects show off a wide range of crystal bead shapes and palettes. Crystals also are commonly combined with other types of beads to highlight their brilliance. Christine Strube joins delicious freshwater pearls with crystals (page 77), while Bonnie Clewans shows off gem-like crystals against dark seed beads (page 65).

Many projects are simple enough for a beginning beader, while others explore using more complex off-loom beading stitches. If you're not familiar with these stitches, the basics section covers them with highly understandable illustrations. In this section, you'll learn about the materials and tools needed to make each project a success, while how-to photography affords you an up-close-and-personal look at some basic techniques. Furthermore, each project in the book is thoroughly explained with clear text and lively illustrations.

This book will open your eyes to the many possibilities of beading with crystals, fueling your creative furnace while providing you with the cool confidence to undertake any project of your choice. And when you don the finished product, well, just watch out! Crystals *always* inspire passion…

Basics

The glistening beads featured in *Beading with Crystals* differ from ordinary glass beads. They're made of leaded glass, a material that's composed of the same ingredients as regular glass (silica, soda, lime, and other compounds) but with an extra kick—a fairly high concentration of lead oxide. The result is a heavier-weight glass with exceptional sparkle. When light hits this type of glass, the lead oxide particles refract in an utterly dazzling way.

The history of leaded crystal beads is an interesting one. The story begins in the 1670s, when Englishman George Ravenscroft discovered a way to outdo the crystal glass being made in Murano, Italy. The Italians were making their crystal using quartz sand and potash, but Ravenscroft added lead oxide to the mix, giving his glass an unequaled brilliance. This new type of glass was relatively easy to cut and engrave, which made it a perfect material for creating exquisite tableware, sculptures, chandeliers, and jewelry.

Up until the late 1800s, crystal items were hand-faceted by skilled artisans. In 1892, Bohemian Daniel Swarovski invented a crystal-cutting machine that made it possible to mass-produce crystal objects. He established his company in the Austrian Tyrol region, where it remains today. As time has passed, Swarovski's methods for cutting crystal have become so refined and high-tech that the facets are especially crisp, and shapes unthinkable of achieving before are common today. Although leaded crystal beads have their origin in and primarily come from Austria, they are also made in other parts of the world, including Egypt.

Crystal beads are available in a wide range of colors. It's quite possible to find just the perfect color and finish for any beaded project. There are bright, jewel-tone colors such as sapphire, ruby, and topaz, as well as subtle, sophisticated hues that resemble denim blue, khaki, deep crimson, and delicate peach. Crystal beads aren't surface dyed, like some beads, but gain their color through chemical elements added during the glass-making stage. Surface treatments are sometimes also added, one of the most popular being aurora borealis (AB), which gives beads a rainbow sheen. When a bead is labeled with a color followed by AB, it will have the treatment on one side of the bead. If it is labeled AB2X, the bead will have been coated on both sides.

Crystal beads also come in a wide variety of shapes. The round and diamond (bicone) shapes familiar to most beaders are just the beginning! Cubes, ovals, rondelles, saucers, drops, and hearts are only some of the shapes waiting to be incorporated into jewelry masterpieces.

When purchasing crystal beads, choose high quality over low price. To qualify as full-leaded crystal, beads need to be composed of at least 24% of lead oxide. The finest crystal beads have 32%. Look for finely cut facets, distinct hues, and high-quality surface treatments. Paying just a bit more per bead will improve the look of a piece tenfold, giving it a truly defined and professional look.

Materials and Tools

If you're making an investment in crystal beads to create a stunning piece of jewelry, take some time to get to know the other supplies you'll need to use to put it together. To get the most professional results, always use the best materials and tools you can afford.

Beads

The sheer number of different types and colors of beads available in today's market is staggering—enough to fill a whole book! So, besides crystal beads, here are the handful that are used in the projects in this book.

Fire-polished beads are glass beads made in the Czech Republic. They start out as machine-cut beads, and are then passed through a flame or tumbled to smooth their surface. Sometimes surface treatments are added for effect. The result is a glistening bead that's not quite as sparkly as crystals.

Freshwater pearl beads are created naturally by freshwater mollusks. They are produced in a variety of colors and shapes, as each kind of species produces a different sort of pearl. The characteristic uneven surface differentiates a freshwater pearl from an ocean one.

Lampworked beads are created by heating the end of a glass rod over a very hot flame until it melts, and then capturing the molten glass onto a thin spinning wire (a mandrel). This type of bead is often made commercially, but many artists make stunning one-of-a-kind lampworked beads.

Lucite beads are made of a strong plastic that can be molded or carved. They come in a wide variety of shapes and can be tinted with any color. Lucite was very popular in jewelry of the 1940s, and it has had a recent resurgence.

Metal beads are simply beads made out of metal. They come in precious and base metal varieties, including sterling silver, gold-filled, silver- and gold-plated (over brass), vermeil (gold over sterling silver), brass, and pewter. They can be machine-stamped, made with molds, and handmade.

Pressed-glass beads are made by pouring molten glass into molds and pressing the molds into shapes. The finest pressed-glass beads often come from the Czech Republic, so these beads are often called "Czech glass." Characteristic shapes include leaves and flowers.

Seed beads are small glass beads made by cutting long, thin tubes of glass into tiny pieces. The most common sizes are between 6° and 14° (largest to smallest) and come in an enormous array of colors.

They are primarily made in the Czech Republic and Japan, and come in three popular types: *Czech seed beads* have a somewhat flat profile, like a donut; *Japanese seed beads* have a taller profile than Czech seed beads and are fairly uniform from bead to bead; and *cylinder beads* have thinner walls than other seed beads and are very uniform in shape.

Stringing Materials

Whether you're doing wirework, stringing, or off-loom beadwork, you need some sort of stringing material to keep the beads together. Here's a list of the most common types.

Beading line was developed by the fishing industry. It's a braided nylon thread that's extremely strong and durable, can be knotted, and is a great choice for working with crystal beads because it doesn't abrade easily. It's generally available in clear, white, moss, and dark gray and comes in 6- to 20-pound test weights. Cut this type of thread with a children's-size household scissors because cutting it with embroidery scissors will dull the blades.

Beading thread is a very pliable thread made of nylon. It's fairly strong and comes in dozens of colors. Some beading threads come pre-waxed, but if yours isn't, coat it liberally with wax or thread conditioner.

Flexible beading wire is primarily used for stringing beads. It's produced by twisting dozens of strands of tiny stainless steel wires together, and then giving it a nylon coating. Secure this type of wire with crimp beads.

Monofilament is a tough, synthetic material developed by the fishing industry. It's available in a clear version that many beaders like, especially for working with crystals. It's stiff enough that you really don't need to use a beading needle while working with it.

Metal wire is used in this book for wireworking projects ranging from head pins to ear wires to brooches. It comes in a range of widths, but 18- to 26-gauge (the smaller the number, the thicker the wire) are the sizes used in this book. You can purchase all types of metal wire, but the most common ones used to make fine beaded jewelry are sterling silver and gold-filled.

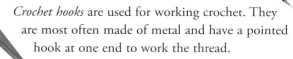

Tools

Having the right tools for the job is a key element to happy beading! The tools are as varied as the bead projects in this book.

Bead stoppers or other types of strong clips are used to keep your beads from sliding off the end of your stringing material while you bead. They are most often used with flexible beading wire.

Beading needles are extremely thin pieces of stiff wire (about the width of a piece of thread) that have a tiny hole on one end and a very sharp point on the other. The most popular for beading include *English beading needles,* which are especially thin and long, and *sharp needles,* which have a stronger body and are somewhat shorter.

Beeswax or thread conditioner, pictured below, is used to prepare thread before stitching. It can help with thread tension and ensure the thread doesn't fray.

Big-eye needles are easy to thread. They are made up of two pieces of extremely thin wire. The wire ends are fused together and then sharpened. To thread it, you pull apart the two unfused wires and pass the thread through the opening.

Chain-nose pliers feature jaws that are flat on the inside but taper to a point on the outside. This type of pliers also comes in a bent version used for grasping hard-to-reach places.

Crimping pliers are used for attaching crimp beads and crimp tubes to beading wire.

Crochet hooks are used for working crochet. They are most often made of metal and have a pointed hook at one end to work the thread.

Embroidery scissors are very sharp scissors with pointed blades. They are used for cutting beading thread.

Emery paper is used in this book for sanding wire smooth.

Flat-nose pliers feature jaws that are flat on the inside and have a square nose.

Jeweler's hammers are used in this book for flattening, hardening, and texturizing wire.

Jeweler's wire cutters have very sharp blades that come to a point. One side of the pliers leaves a cut, while the other side leaves a flat, or "flush," cut.

Jigs, such as pictured below, are made up of a flat board with pegs. The pegs are placed at desired intervals to help bend wire into perfect loops. You can purchase commercial jigs or make your own.

Metal hand files, or needle files, have very fine teeth. They are used in this book for smoothing out wire ends or sharpening wires to a point.

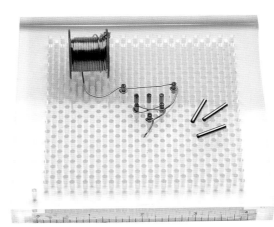

Findings

In beading, you'll often need more than just beads and string to create your final piece. Findings are that extra ingredient. They are usually made of metal and are meant to connect, finish, and embellish your jewelry designs.

Chain, pictured left, is made up of connected loops of wire. The loops can come in several forms, including round, oval, twisted, and hammered.

Charms are small pieces of metal with a hole or loop from which you can hang it from something else, such as a chain. They are sometimes made with lively, detailed images or may simply include words, symbols, or decorative designs.

Clasps connect wire ends to keep a necklace or bracelet in place. There are dozens of different types. Here are some of the most common.

> *Box clasps* have one half that's comprised of a hollow box. The other half is a tab that clicks into the box to lock the clasp.
>
> *Hook-and-eye clasps* have one half that's shaped like a hook, the other half like a loop, or "eye." The hook passes through the eye to secure the clasp.
>
> *Lobster-claw clasps* are spring-activated clasps that are shaped like their name.
>
> *Magnetic clasps* use powerful magnets to make the connection between one half of the clasp and the other. Use these only with fairly lightweight pieces, and if you have a pacemaker, don't use them at all.
>
> *Toggle clasps* have one half that looks like a ring with a loop attached to it, and the other half

wire, the pliers won't mar or dent it.

Round-nose pliers feature cylindrical jaws that taper to a very fine point.

Steel blocks are thick, smooth chunks of steel upon which you can hammer wire.

Thread burners or lighters help hide the clipped ends of synthetic threads by melting them into a tiny ball. Thread burners are very precise—they have a wire end that, once warmed up, you can touch to the end of the thread and the thread melts away. A lighter does the same job, but it isn't as precise.

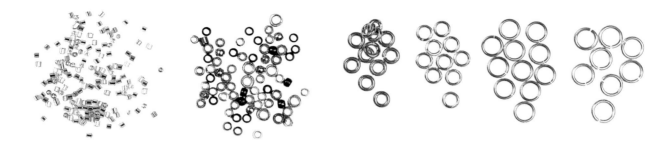

looks like a bar. Pass the bar through the ring, and once the bar lays parallel on top of the ring, you've secured the clasp.

Connectors allow you to make the transition from one beaded strand to many strands. In this book, connectors are used as earring components.

Crimp tubes and crimp beads, pictured above, are used to connect beading wire to a finding.

Ear wires, pictured below, are the findings you use to make pierced earrings. They include a jump ring–like loop onto which you can add an earring dangle. They come in different shapes, including *French ear wires,* which look like upside-down U shapes; *hoops,* which are simply rounds of thin wire with a catch to hold them in place on the ear; and *lever backs,* which are much like French ear wires, but have a safety catch on the back to hold the earring in place.

Eye pins are used in this book to make beaded links. They are made with straight pieces of wire with a simple loop at one end.

Head pins, pictured below, are used for stringing beads to make dangles. Simple head pins are composed of a straight wire with a tiny disk at one end to hold beads in place.

Jump rings, pictured above, are circular loops of wire used to connect beadwork to findings or findings to findings. They come in open and soldered-closed versions. Always open a jump ring with two pairs of pliers, one positioned on each side of the split. Push one pair of pliers away from you, and pull the other one toward you. This way the ring will be opened laterally, instead of horizontally (figure 1), which can weaken the wire. (You'll also open any other wire loop this way.)

fig. 1

Pin backs, pictured below, have a flat front with several holes through which you can secure it to a piece of beadwork. The back of the pin has a sharp wire with a lever catch to keep it closed.

Beading Techniques

There's a wide variety of skills you'll need to know to make all the projects in this book. But don't fret. Just study the how-to information below and you'll be beading like a pro.

Stringing

Stringing beads is a simple act—simply pass the thread or wire through a bead, and you've got it! It's how you arrange beads on the stringing material that creates masterpieces—that's what takes practice.

Crimping wire is a technique used to attach wire to a finding (like a clasp). Start by stringing 1 crimp bead and the finding. Pass the wire end back through the crimp bead in the opposite direction. Next, slide the crimp bead against the finding so it's snug, but not so tight that the wire can't move freely. Squeeze the crimp bead with the back U-shaped notch in a pair of crimping pliers (photo 1). Turn the crimp bead at a 90° angle and nestle it into the front notch. Gently squeeze the bead so it collapses on itself into a nicely-shaped tube (photo 2).

Knotting

Knowing how to tie knots is very important if you're working off-loom beading, but the skill will also come in handy for other beading techniques.

Make an *overhand knot* by forming a loop with the thread, passing the thread end through the loop, and pulling tight (figure 2).

fig. 2

Make a *square knot* by first forming an overhand knot (above), right end over left end, and finish with another overhand knot, this time left end over right end (figure 3).

fig. 3

A *surgeon's knot* is an extremely secure version of a square knot. It's basically made the same way as the square knot, but when you make your first overhand knot, wrap the thread around itself a few times before passing it through the loop. Finish the knot with another overhand knot and pull tight (figure 4).

fig. 4

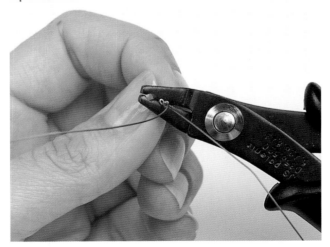
photo 1

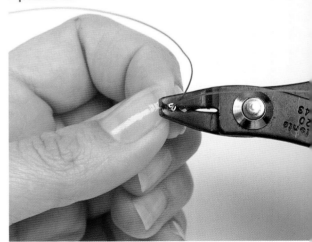
photo 2

Off-Loom Beading

Off-loom beadwork is produced by connecting beads with a needle and thread (and without a loom—thus the name). The result is a supple fabric of beads. There are dozens of off-loom beading stitches, and each stitch has any number of tricks that go along with it. But the following section will get you up and running to make the off-loom projects in this book.

EDGINGS

Beaded edging is a way to add extra pizzazz to edges. Two techniques are used in this book.

Simple fringe is made by stringing on a length of beads and skipping the last bead strung before passing the needle back through the rest of the strung beads (figure 5).

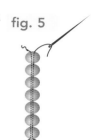

fig. 5

Simple edging starts by exiting the thread through a bead at the edge the beadwork where the beads sit side by side.

General Off-Loom Beading Terms and Techniques

There are several general terms and techniques that are basic to off-loom beading. Here's a list of the few you'll need to know to make the off-loom pieces in this book.

Pass through means you'll pass the needle through the beads in the same direction as they were strung. *Pass back through* means you'll go through in the opposite direction.

You've made a *row* of off-loom beadwork when you've stitched beads in a line back and forth, and it results in a flat piece of beadwork. A *round* is created when you've stitched beads in circles, creating circular or tubular pieces of beadwork.

A *stop*, or *tension*, bead is used at the end of a working thread to keep beads from slipping off. To make one, simply string a bead and pass through it again once or twice. Once you've worked your piece enough that the beads are secure, you can easily remove the stop bead.

The *tail thread* is the length at the end of the thread that remains below the first bead you strung. The *working thread* is the portion of thread between the needle and the first beads strung. You use it to do your stitching.

To *end a thread*, weave through several beads on the body of the beadwork, tie an overhand knot on the threads between beads, pass through a few more beads to hide the knot, and trim the knot close to the work. You also use this technique to *secure the thread* when you're finishing a piece.

To *start a new thread*, thread a needle with the required length of thread. Pass through several beads on the body of the beadwork, tie an overhand knot between beads as desired, and continue to weave through the beads until you exit from a place where you can keep stitching.

Weaving through beads on an off-loom piece of beadwork means you're passing the needle through beads on the body of the beadwork so you can exit elsewhere. Keep your thread hidden by passing only through adjacent beads.

To *reinforce* off-loom beadwork, simply pass through the stitched beads more times than is required. This stiffens and strengthens the work.

SIMPLE EDGING

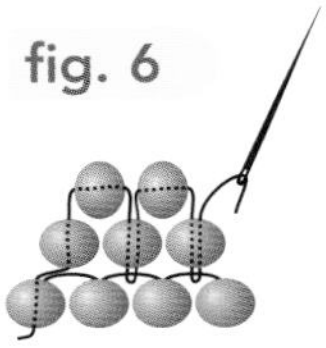

1. String 3 beads and loop under the exposed thread between the edge bead you last exited and the adjacent one.

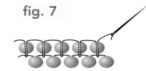

fig. 6

2. Pass back through the third bead just strung. String 2 beads, and loop under the thread between the next 2 edge beads. Repeat to add 2 beads at a time along the base edge (figure 6).

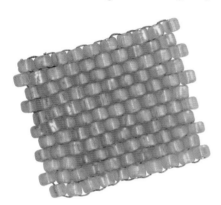

BRICK STITCH

Brick stitch creates a smooth beaded fabric with a brick-like pattern.

fig. 7

1. Begin by creating a base row of ladder stitch (see next page).

2. Pass up through the last bead on the base and string 2 beads. Loop under the exposed thread between the next 2 base beads, and pass back through the second bead just strung. For the following stitches, string 1 bead, and loop under the thread between the next 2 base beads. Begin the next row as you did the first one (figure 7).

fig. 8

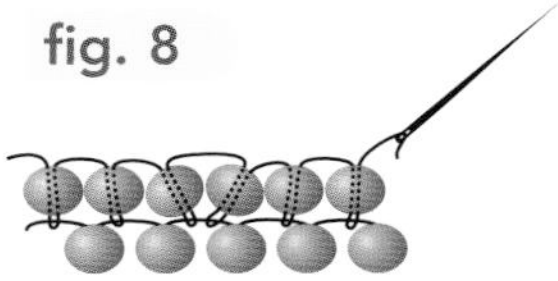

Make a *brick-stitch increase* by stitching 2 beads—one at a time—where one bead would go (figure 8). When you come to the increase in the following row, work across as you normally would.

In this book, brick stitch is done in rounds. To work *circular brick stitch,* start by creating a circular base. Work

brick stitch around the base. When you've finished the round, position your needle for the next round, making a "step up," by passing down through the first bead added in the round, and up through the second bead in the round (figure 9).

fig. 9

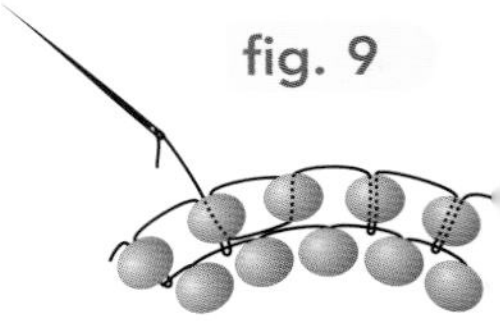

HERRINGBONE STITCH

Herringbone stitch produces beadwork with a chevron, or herringbone, pattern.

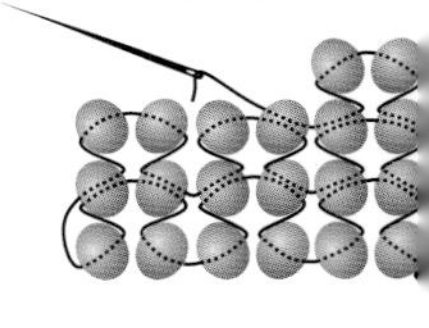

1. Start with a base row of ladder stitch (see next page) that has an even count of beads.

fig. 10

2. Exit up through a bead at the end of the base round, string 2 beads, and then pass down through the next bead on the base row and up through the following base-row bead. Repeat across the row.

3. Position your needle for the next round, making a "step up," by passing back through the last bead strung. Work subsequent rows as you did this one, stringing 2 beads, then passing down through the next bead added in the previous row and up through the following one (figure 10).

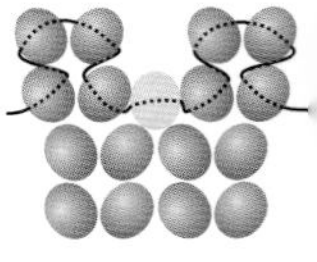

Make a *herringbone stitch increase* by first working the 2-bead stitch as usual. Then pass down through the next bead on the previous round, string 1 or more beads, and pass up through the next bead on the previous round (figure 11).

fig. 11

LADDER STITCH

Ladder stitch is often used to make a foundation row for brick or herringbone stitch.

1. Begin by stringing 2 beads. Pass through the beads again to make a circle and manipulate them so they sit side by side (figure 12).

fig. 12

2. String 1 bead and pass down through the second bead initially strung and up through the one just strung. Repeat to add 1 bead at a time until you reach the desired length (figure 13).

fig. 13

3. String 3 dark beads, 1 light bead, and 3 dark beads. Pass back through the first light bead added to the previous vertical row. String 3 dark beads and pass back through the next light bead on the base row. This makes the second vertical row.

4. String 3 dark beads, 1 light bead, and 3 dark beads. Pass through the light bead added to the previous vertical row. String 3 dark beads, 1 light bead, and 3 dark beads. Pass back through the last light bead just strung. Repeat steps 3 and 4 across the base row (figure 14).

fig. 14

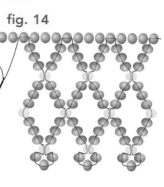

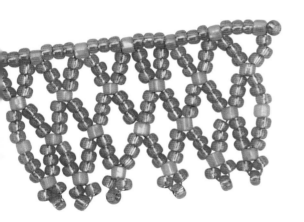

NETTING

Netting beads makes an open, airy type of beadwork that resembles a fishnet. The type of netting used in this book is called *vertical netting*.

1. Make a stop bead (page 15), then create a base row by stringing a sequence of 3 dark beads and 1 light bead as many times as necessary to reach your desired width.

2. String 3 dark beads, 1 light bead, 3 dark beads, 1 light bead, 3 dark beads, 1 light bead, and 3 dark beads. Then pass back through the last light bead just strung. This is your first vertical row.

PEYOTE STITCH

Peyote stitch forms a supple type of beadwork with a staggered bead pattern. There are several peyote-stitch techniques used in this book.

TO WORK A FLAT, EVEN-COUNT PEYOTE STITCH

1. String an even number of beads (this strand will make up your first and second rows).

2. String 1 bead and pass through the second-to-last bead initially strung. Skip 1 bead and pass through the next bead on the initial strand. Continue across to the end to complete the third row.

3. Adjust the beads so they look like a spine, and tie a knot with the tail and the working thread to hold the beads in place.

fig. 15

4. To make the next row, simply string 1 bead, pass through the last bead added in the previous row, and continue across, adding 1 bead between each bead added on the previous row (figure 15). *Note:* Count peyote-stitched rows at a diagonal, as shown in the sample (figure 16).

fig. 16

TO WORK FLAT, ODD-COUNT PEYOTE STITCH

1. String an uneven number of beads (this strand will make up your first and second rows).

2. Begin your third row by stringing 1 bead and passing through the second-to-last bead initially strung. Skip 1 bead and pass the needle through the next bead on the initial strand. Continue across. When you come to the end of the row, you will end up with 2 beads sitting side by side. After you add the last bead, pass the needle through the adjacent bead, and weave through the beads to exit from the last bead you added.

3. Work the fourth row in regular peyote stitch. Work the fifth row as you did the third row, either weaving through the beads to make the turnaround, or by simply passing under the exposed loop of thread between the last two rows, passing through the last bead added, and starting the next row (figure 17).

fig. 17

TO ZIP UP A PIECE OF FLAT PEYOTE BEADWORK

Zipping up a piece of flat peyote-stitched beadwork means you'll be connecting the last row you stitched either to another piece of beadwork, or to the first row of the same piece so you can make a tube.

1. Make sure the beads on the rows you're connecting interlock like a zipper (thus the name of this technique). If they don't, add or subtract 1 row to one of the rows.

2. Weave through the beads to exit from an end bead of the previous row and pass the needle through the corresponding bead on the second piece. Pass through the second-to-last bead added on the first piece of beadwork and then the corresponding bead on the second piece. Continue across, lacing the rows from each piece of beadwork together (figure 18). The result should look like one continuous piece of beadwork.

fig. 18

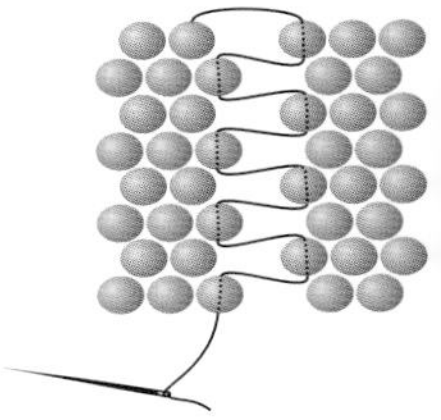

A few projects in this book call for *peyote-stitched decrease* within a row or round. To do so, simply skip the bead you might normally add in a stitch and pass through the next bead on the previous row. Pull tight. When you work the next row, add 1 bead over the decrease and continue across. Work the following row as you would a regular row (figure 19).

fig. 19

TO MAKE A SCULPTURAL TUBE

For a *tubular peyote stitch*, you work in rounds to create sculptural tube.

1. String on the specified amount of beads, then use a square knot to tie the beads into a circle.

2. Pass through the first bead strung to hide the knot. Then string 1 bead, skip 1 bead, and pass through the next bead. Pull tight.

3. Continue around and position your needle for the next round, making the "step up" by passing through the first bead on the foundation circle and the first bead on the current round (figure 20).

fig. 20

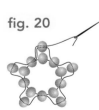

4. Work the following round by stringing 1 bead and passing through the next bead on the previous round. *Note:* You'll step up after each round, passing the needle through the first beads on the previous and current rounds. Many beaders find it helpful to begin tubular peyote stitch around a form, such as a dowel or a chopstick (figure 21).

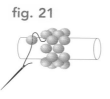
fig. 21

QUARE STITCH

quare stitch produces a relatively stiff beaded fabric in which all the beads sit side by side to create a grid.

. String a base row of beads long enough to make up your first row. Begin the second row by stringing 1 bead and passing through the last bead strung on the previous row and the bead just strung. The two beads just worked should sit side by side.

. String 1 bead and pass through the next bead on the base row and the bead just strung. Continue across the base strand, stitching 1 bead to 1 bead until you reach the end (figure 22).

fig. 22

. Start the next row by stringing 1 bead and stitching it to the last bead added on the previous row (as you did with the first row). Working in the opposite direction from the first row, continue across, stitching 1 bead to 1 bead (figure 23).

fig. 23

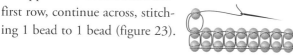

RIGHT-ANGLE WEAVE

Right-angle weave creates an open weave of fabric made up of little squares, which are called "units" in this book. The technique can be done using either 1 or 2 needles.

TO WORK SINGLE-NEEDLE RIGHT-ANGLE WEAVE

1. String 4 beads and use a square knot to tie the beads into a tight circle. Pass through the first 3 beads strung, then string 3 beads and pass through the bead just exited and the first 2 beads just added. Repeat, adding 3 beads at a time in a figure eight–like fashion, until you reach the desired row length (figure 24).

fig. 24

2. Start the next row by exiting up through the first bead positioned along the side of the first row. String 3 beads and pass up through the side bead just exited, through the 3 beads just strung, and down through the next side bead on the previous row (figure 25).

fig. 25

3. String 2 beads and pass through the third bead added in the previous unit, down through the side bead just exited, and through the first bead just strung (figure 26).

fig. 26

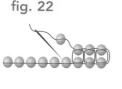

4. String 2 beads and pass up through the next side bead on the previous row, through the first bead added in the previous unit, through the 2 beads just strung, and down through the next side bead on the previous row (figure 27). Continue working the row this way until you reach the end. Start the new row as you did this one.

fig. 27

TO WORK DOUBLE-NEEDLE RIGHT ANGLE WEAVE

1. Thread 1 needle on each end of a length of thread. String 4 beads onto the thread and slide them to the middle, then cross the right needle through the last bead strung. String 3 beads on the left needle, then cross the right needle through the last bead strung. Repeat, adding 3 beads at a time, until you reach the desired row length (figure 28).

fig. 28

2. Turn the work so the first unit on the first row is on top. Weave the needles through the beads so they exit in opposite directions from the first bead positioned along the side of the first row. String 3 beads on the left needle and pass through the next side bead on the previous row. Cross the right needle through the third bead just strung (figure 29).

fig. 29

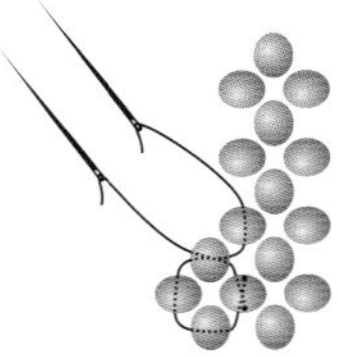

3. String 2 beads on the left needle and pass through the next side bead on the previous row. Cross the right needle through the second bead just strung (figure 30). Continue working the row this way until you reach the end. Start the new row as you did this one.

fig. 30

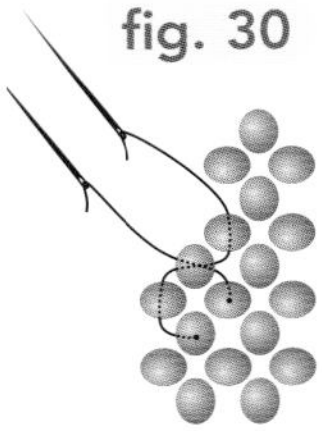

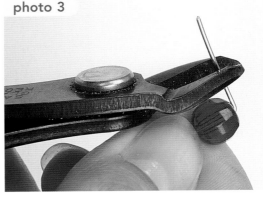

photo 3

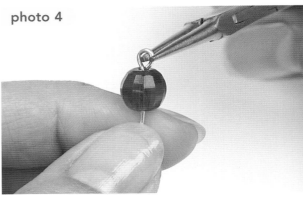

photo 4

Wireworking

Bending, shaping, and finishing wire are basic technique used in many of the beautiful pieces in this book.

Filing and sanding are necessary for smoothing rough wire ends. Use a flat metal file or emery paper to achieve this, touching the wire occasionally as you go to check for any rough spots. You can also file the end of a wire t create a sharp point, as for a pin stem. Do so by using a flat file to file the wire at a 45° angle on all sides, makin a sharp point.

Flush cutting wire involves using the flat, or flush, side o the wire cutters to make the cut so the wire end is flat.

Wire loops come in two versions, simple and wrapped.

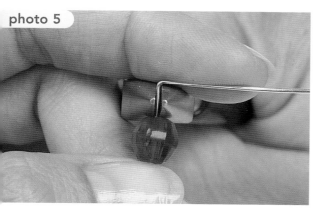

photo 5

MAKING A SIMPLE LOOP

1. Use chain-nose pliers to make a 90° bend ⅜ inch from the end of the wire; or, if you're using the loop to secure a bead (as with a bead dangle), and make the 90° bend right at the top of the bead. Cut the wire ⅜ inch from the top of the bead (photo 3).

2. Use round-nose pliers to grasp the wire end and roll the pliers until the wire touches the 90° bend (photo 4).

MAKING A WRAPPED LOOP

1. Use chain-nose pliers to make a 90° bend in the wire 2 inches from one wire end or ¼ inch from the top of a bead (photo 5). Then use round-nose pliers to grasp the bend, shape the wire over the top jaw (photo 6), and swing it underneath to form a partial loop (photo 7).

2. Use chain-nose pliers or your fingers to wrap the wire in a tight coil down the stem (photo 8). Then trim the excess wire close to the wrap and use chain-nose pliers to tighten the wire end.

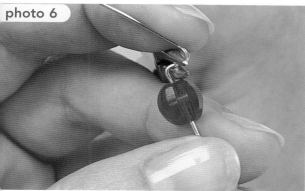

photo 6

Wrapping, or coiling, wire is primarily used in this book both for attaching one wire to another and for creating decorative coils. Start by grasping the base wire tightly in one hand. Hold the wrapping wire with your other hand and make one wrap. Reposition your hands so you can continue to wrap the wire around the base wire, making tight revolutions (photo 9).

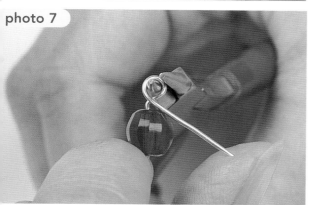

photo 7

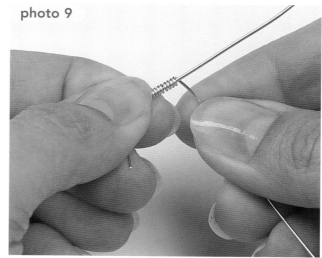

photo 9

photo 8

Projects

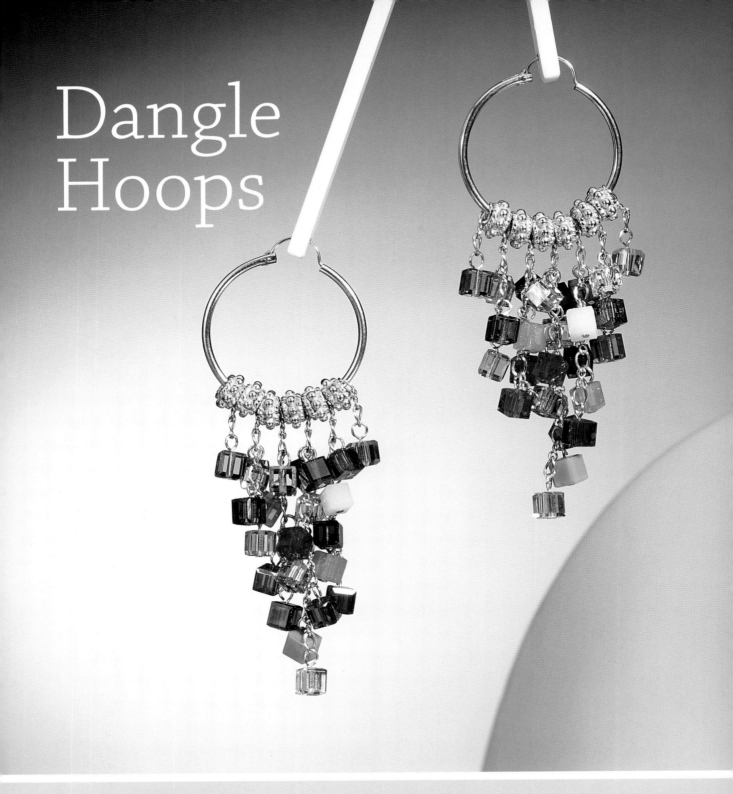

Dangle
Hoops

Brightly colored crystal cubes cascade from round hoops, creating vivid visual impact.

Designer: Marlynn McNutt

Materials

50 assorted color 4 mm crystal cube beads

50 gold-filled 1-inch head pins

12 vermeil spacer beads

2 gold-filled 1-inch ear hoops

4 inches of gold-filled flat figure-8 chain

Tools

Wire cutters

Round-nose pliers

Chain-nose pliers

Techniques

Simple loop (page 21)

Instructions

1. Cut 2 pieces of chain with 2 large links, one at each end. **fig. 1** Cut 2 pieces with 4 large links, one at each end and 2 in the middle. Cut 2 pieces with 6 large links, one at each end and 4 in the middle. And cut 1 piece with 8 large links, one at each end and 6 in the middle (figure 1). Set aside.

2. Use 1 head pin to string 1 cube bead. Make a simple loop to secure the bead and trim any excess wire. Repeat to make 25 cube dangles in all. Set aside.

3. Attach 1 dangle to an end link of one of the 2-link chains. Repeat for the other 2-link chain. Set aside.

4. Attach 1 dangle to each large link of one of the **fig. 2** 4-link chains, leaving the top link open. You'll add 3 dangles in all (figure 2). Repeat for the other 4-link chain. Set aside.

5. Attach 1 dangle to each large link of one of the 6-link chains, leaving the top link open. You'll add 5 dangles in all. Repeat for the other 6-link chain. Set aside.

6. Attach 1 dangle to each large link of the 8-link chain, leaving the top loop on the chain open to attach to the ear hoop. You'll add 7 dangles in all.

7. Open the catch on one of the earring hoops. String the end link of one of the embellished 2-link chains onto the hoop. String 1 spacer. String 1 embellished 4-link chain and 1 spacer. String 1 embellished 6-link chain and 1 spacer. String the 8-link chain. Repeat in reverse order with the remaining chains.

8. Close the ear hoop.

9. Repeat steps 1 through 8 to make the second earring.

Spirals

Wire spirals adorned with coils, crystals, and spacer beads
are assembled into an exciting necklace and earring set.

Making the Earrings

Designer: Wendy Witchner

Materials for Earrings

4 assorted color 6 mm crystal bicone beads

12 assorted color 4 mm crystal bicone beads

2 sterling silver 6 mm Bali-style rondelle beads

2 sterling silver 5 mm Bali-style bead caps

4 sterling silver 5 mm Bali-style spacer beads

4 sterling silver 3.5 mm Bali-style spacer beads

2 sterling silver 1-inch paddle head pins

1 pair of sterling silver ear wires

36 inches of 20-gauge dead-soft sterling silver wire

36 inches of 24-gauge dead-soft sterling silver wire

Liver of sulfur

Clear adhesive cement (optional)

Tools

Wire cutters

Chain-nose pliers

Round-nose pliers

Wire brush

½-inch-diameter dowel

Techniques

Coiling wire (page 21)

Simple loop (page 21)

Opening and closing loops (page 13)

Instructions

Making the Coils

1. Cut a 10-inch length of the 20-gauge wire. Tightly coil the 24-gauge wire down the length of the 20-gauge wire. Trim the tails close to the coil.

2. Dip the coiled wire into a liver of sulfur solution, rinse, and brush.

3. Slide the coil off the end of the 20-gauge wire and trim off a ½-inch piece. Repeat to make 5 more ½-inch coils. Set aside.

Making the Center Link

1. Cut a 2-inch piece of the 20-gauge wire. Make a simple loop at one end.

2. **fig. 1**  String 1 coil, one 6 mm crystal bead, and 1 coil. Trim the 20-gauge wire ⅜ inch from the last coil strung and form a simple loop (figure 1). Set the center link aside.

Making the Spiral

1. Cut a 6-inch length of 20-gauge wire. String 1 coil, one 4 mm bicone bead, one 5 mm spacer bead, one 4 mm bicone bead, 1 coil, one 4 mm bicone bead, one 6 mm rondelle bead, one 4 mm bicone bead, 1 coil, one 4 mm bicone bead, one 5 mm spacer bead, one 4 mm bicone bead, and 1 coil.

2. Keeping the beads and coils in place with your fingers, wrap the beaded wire around the dowel for 3 rotations. Use your fingers to adjust the coil so all the beads are lined up down its center (figure 2). Set aside.

fig. 2

Assembling the Earrings

1. Slide the center link through the middle of the spiral.

2. fig. 3

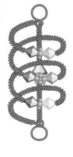

Tightly wrap the exposed 20-gauge wire at the top of the spiral to the base of the simple loop at the top of the link. Repeat for the other end of the spiral and link (figure 3). If desired, dab clear adhesive cement on the top and bottom wraps.

3. Use 1 head pin to string one 3.5 mm spacer bead, one 6 mm crystal bead, 1 bead cap from inside to outside, and one 3.5 mm spacer bead. Cut the wire to ⅜ inch above the last bead strung and form a simple loop. Connect the dangle to the bottom loop of the center link.

4. Connect the ear wire to the top loop of the center link.

5. Repeat all steps to make the second earring.

Making the Necklace

Materials for Necklace

5 assorted color 6 mm crystal bicone beads

50 assorted color 4 mm crystal bicone beads

6 sterling silver 7 mm Bali-style spacer beads

4 sterling silver 4 mm Bali-style spacer beads

Assortment of 15 sterling silver 5 mm Bali-style spacer beads

2 sterling silver ½-inch clasp hooks

5¼ inches of 3 x 6 mm sterling silver oval link chain

80 inches of 20-gauge sterling silver wire

150 inches of 24-gauge sterling silver wire

Tools

Wire cutters

Chain-nose pliers

Round-nose pliers

⅜-inch-diameter dowel

Instructions

1. Follow the first 3 sections for the earrings (page 27) to make 5 spiral components. This time make the coils only ½ inch in diameter, and don't include the end dangle. Set the components aside.

2. Cut a 1-inch length of 20-gauge wire and make a simple loop on one end. String one 4 mm bicone bead, one 4 mm spacer bead, and one 4 mm bicone bead. Make a simple loop to secure the bead. Repeat 3 times to make 4 small beaded links in all. Set aside.

3. Cut a 1-inch length of 20-gauge wire and make a simple loop on one end. String one 4 mm bicone bead, one 7 mm spacer bead, and one 4 mm bicone bead. Make a simple loop to secure the bead. Repeat 5 times to make 6 large beaded links in all. Set aside.

4. Cut the chain into three 1¼-inch pieces and two ¾ inch pieces. Use the small beaded links to connect the chain pieces.

5. Use the large beaded links to connect the spiral components.

6. Use the clasp hooks to connect the chain to the spiral section of the necklace.

Violets

This simple-to-assemble set shows off the multi-faceted appeal of crystals.

Making the Bracelet

Designer: Val Hirata

Materials for Bracelet

33 lilac AB 4 mm bicone beads

34 violet AB 4 mm bicone beads

21 amethyst 3.5 mm crystal montées

2 sterling silver 1 x 2 mm crimp tubes

1 sterling silver 6 mm soldered jump ring

9 mm sterling silver lobster clasp

34 inches of .010 flexible beading wire

Tape

Tools

Wire cutters

Chain-nose or crimping pliers

Techniques

Stringing (page 14)

Crimping (page 14)

Note

The materials are for a 7¼-inch bracelet. Add or subtract beads for a larger or smaller size.

Instructions

1. Cut the beading wire into one 14-inch piece and one 20-inch piece. Set aside.

2. Tape the two pieces of wire to the work surface, about 2 inches from the ends and about ½ inch apart.

3. On the short wire, string a sequence of 1 lilac bead, 1 montée, 1 violet bead, and 1 montée 10 times. String 1 lilac bead, 1 montée, and 2 violet beads. Tape the wire ends to the work surface.

4. On the long wire, string 2 lilac beads and pass through the first montée added to the short wire. The wires should cross through the montée. String 2 violet beads and pass through the next montée added to the short wire. String 2 lilac beads and pass through the next montée added to the short wire (figure 1). Repeat down the length of the short wire until you've passed through all the montées. Exit from the last bead strung on the short wire.

fig. 1

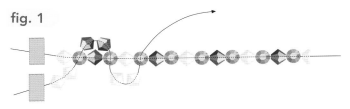

5. Pair the wire ends and, holding the wires tight to maintain the tension, string 1 crimp tube and the soldered jump ring. Pass back through the crimp tube and the last lilac bead from the short wire (figure 2). Squeeze it tight with chain-nose pliers, or crimp it with crimping pliers. Trim the excess wire close to the bead.

fig. 2

6. Undo the tape at the other end of the bracelet and repeat step 5 with the lobster clasp.

Making the Necklace

Materials for Necklace

2 lilac 6 mm crystal bicone beads

68 violet AB 4 mm crystal bicone beads

180 lilac AB 4 mm crystal bicone beads

42 amethyst 3.5 mm crystal rose montées

2 lilac AB 4 mm crystal round beads

8 sterling silver 1 x 2 mm crimp tubes

2 sterling silver 7 mm soldered jump rings

2 sterling silver 7 mm open jump rings

2 sterling silver 1-inch head pins

2 sterling silver one-to-three loop connectors

9 mm sterling silver lobster clasp

80 inches of .010 flexible beading wire

48 inches of .014 flexible beading wire

Tape

Tools

Wire cutters

Chain-nose or crimping pliers

Techniques

Stringing (page 14)

Crimping (page 14)

Instructions

Making the Centerpiece

1. Follow the instructions on page 31 to make 2 bracelets, but don't finish the ends.

2. Lay the bracelets next to each other on the work surface with the remaining 12-inch piece of .010 beading wire between them. Use a piece of tape to secure the ends of all the wires to the table. String 1 lilac 4 mm bead and pass through the first bicone bead between the montées on one of the bracelets. String 1 lilac bead and pass through the next bicone bead between the montées on the opposite bracelet. Repeat down the bracelet, lacing the two bracelets together (figure 1). When you reach the end, pull th wires tight to create a subtle ruffle.

   fig. 1

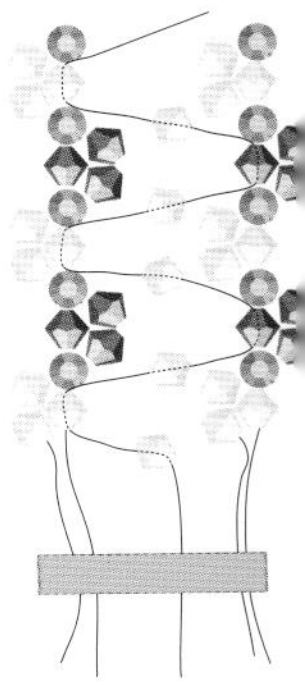

3. Gather the ends of all 5 wires and string 1 crimp tub and 1 soldered jump ring. Pass back through the crimp tube and pull tight. Use chain-nose pliers to squeeze the tube flat or use crimping pliers to crimp the tube. Repeat for the other wire ends.

Adding the Straps

1. Cut a 24-inch piece of the .014 beading wire and pass it through the soldered jump ring you last added. Slide the jump ring to the center of the wire. Pair the wire ends and string 1 crimp tube and 1 lilac 6 mm bead. Snug the crimp against the ring and squeeze the tube flat or crimp it (figure 2).

fig. 2

Separate the wire ends and string 28 violet beads, 1 crimp tube, and the first loop on the 3-loop side of one of the connectors. Pass back through the tube, snug the beads, and squeeze the tube flat or crimp it. Repeat for the other wire end, this time connecting to the third loop on the connector.

fig. 3

Use 1 head pin to string 1 round bead. Make a simple loop to secure the bead. Connect the bead dangle to the second loop on the connector (figure 3).

4. Repeat steps 1 through 3 to create a strap on the other side of the necklace.

5. Attach 1 jump ring to the remaining loop on each connector. Attach the lobster clasp to one of the jump rings.

Vintage

These lively earrings and fantastic ring were designed to showcase unusual vintage crystal beads.

Making the Ring

Designer: Katherine Song

Materials for Ring

5 blue/green 12 mm vertically drilled vintage crystal pagoda-style beads

3 black 4 mm crystal round beads

15 burgundy 3 mm crystal bicone beads

30 inches of 24-gauge silver-colored art wire

10 inches of 28-gauge silver-colored art wire

Tools

Wire cutters

Flat-nose pliers

Round-nose pliers

Ring mandrel or finger-wide dowel

Techniques

Simple loop (page 21)

Coiling wire (page 21)

Instructions

1. Place the ring mandrel in the center of the 24-gauge wire. Wrap the wire around the mandrel 5 to 7 times to make a ring base. Leave 10 inches of unwrapped wire at each end.

2. String 1 black bead onto 1 wire end and cross through it with the other wire end. The bead should rest on the outside of the base. Wrap each wire end around the base on each side of the bead (figure 1).

 fig. 1

3. String 1 vintage bead so it sits ¼ inch from the ring base. Draw the wire down the back of the vintage bead. Hold the wire and bead and twist 2 or 3 times (figure 2).

fig. 2

4. Repeat step 3 to add the other 4 vintage beads, forming a flower shape (figure 3).

fig. 3

5. Wrap one wire end to the ring base, just underneath one side of the flower. String 1 black bead and make a tiny simple loop to secure it. Trim any excess wire. Repeat with the other wire end on the other side of the flower.

6. Secure one end of the 28-gauge wire to the middle of flower.

7. String 5 burgundy beads on one end. Wrap the wire around the middle of the ring base to make 1 stamen. Repeat twice to make 3 stamens in all.

8. Wrap the remaining wire 3 times along the ring band, underneath the flower (figure 4). Coil the wire around one of the black beads and trim the wire close to the coil.

fig. 4

Making the Earrings

Materials for Earrings

2 fuchsia 14 mm top-drilled (front to back) crystal flower beads

2 light violet 14 mm top-drilled (front to back) crystal flower beads

2 blue/green 12 mm vertically drilled vintage crystal pagoda-style beads

4 clear AB 12 mm top-drilled (front to back) crystal round disk beads with faceted edges

2 topaz 8 mm top-drilled crystal bicone beads

2 fuchsia 6 mm crystal cube beads

4 opaque light pink 4 mm crystal round beads

2 olive 4 mm crystal bicone beads

2 topaz 5 mm crystal-studded rondelle spacer beads

8 sterling silver 4 mm oval jump rings

4 sterling silver 2 x 2 mm crimp tubes

10 sterling silver 1 x 2 mm crimp tubes

2 sterling silver clip earring findings

24 inches of .019 beading wire

Tools

Wire cutters

Chain-nose pliers

Techniques

Stringing (page 14)

Instructions

1. Cut 2 inches of beading wire.

2. Use the wire to string one 1 x 2 mm crimp tube. Place it close to the end of the wire and use chain-nose pliers to flatten the tube.

3. fig. 1 String 1 pink round bead, 1 cube bead, 1 pink round bead, and one 2 x 2 mm crimp tube. Slide the tube so it's about 1¼ inches from the crimp tube placed in step 2. Pass back through the crimp tube just strung to make a 4 mm loop. String one 1 x 2 mm crimp tube and 1 light violet flower bead. Pass back through both of the open crimp tubes, leaving a 6 mm loop around the flower. Flatten the tubes and trim any excess wire (figure 1). This is dangle A. Set aside.

4. fig. 2 Cut 2 inches of wire. String 1 fuchsia flower bead and slide it to the middle of the wire. Fold the wire in half. Pair the ends and string 1 rondelle spacer bead and one 1 x 2 mm crimp tube. Separate the ends and pass one wire end back through the crimp tube to leave a 4 mm loop at the end. Flatten the crimp tube and trim any excess wire (figure 2). Set aside.

5. Cut 7 inches of wire. Fold it about one-third of the way down. Pair the wire ends and string one 2 x 2 mm crimp tube. Slide the tubes up the wire until you have a 4 mm loop at the end. Flatten the tube.

6. fig. 3  On the left wire end, string 1 topaz bead and one 1 x 2 mm crimp tube. On the right wire end, string 1 pagoda bead and the dangle made in step 4 (figure 3).

7. fig. 4  Pass the right wire end back through the crimp tube and the topaz bead. Continue to loop the wire around to make a ¾-inch circle, and pass through the pagoda bead and the dangle again. Flatten the crimp tube and trim the left wire close to the crimp tube. On the right wire, string 1 olive bicone bead and one 1 x 2 mm crimp tube. Flatten the tube and trim any excess wire (figure 4).

8. Attach 1 jump ring to 1 disk bead. Connect this dangle to the piece made in step 6, between the topaz and pagoda beads. This completes dangle B. Set aside.

9. Connect 3 jump rings. On one end, attach 1 disk bead and the earring finding. On the other end, attach the top loops of dangles A and B.

10. Repeat steps 1 through 9 to make the second earring.

Superstar

You're certain to be the center of
attention when you step out wearing
this star-studded set!

Making the Necklace

Designer: Karli Sullivan

Materials for Necklace

70 black 6 mm top-drilled crystal bicone beads

6 fuchsia 8 mm crystal cube beads

14 black 4 mm crystal cube beads

28 black AB2X 4 mm crystal bicone beads

28 fuchsia AB2X 4 mm crystal bicone beads

24 black 3 mm crystal bicone beads

48 turquoise size 11° seed beads

4 grams of fuchsia size 15° seed beads

54 sterling silver 4 mm Bali-style spacer beads

76 sterling silver 3 mm Bali-style spacer beads

2 sterling silver crimp tubes

2 sterling silver crimp bead covers

1 sterling silver toggle clasp

6-pound test beading line

24 inches of .015 flexible beading wire

50 inches of .010 flexible beading wire

Tools

Scissors

2 big-eye needles

Thread burner or lighter

Bead stop or other clip

Crimping pliers

Wire cutters

Chain-nose pliers

Techniques

Right-angle weave (page 19)

Square knot (page 19)

Stringing (page 14)

Instructions

Making the Stars

1. Thread 1 needle on each end of a 1-foot length of line.

2. Use top-drilled beads to work a strip of double-needle right-angle weave 3 units long (figure 1).

fig. 1

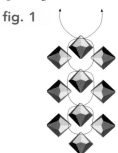

3. Weave the first and last beads together to make a ball (figure 2).

fig. 2

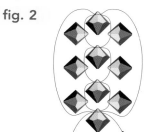

4. Use a square knot to tie the thread ends together. Carefully burn the thread ends. Set aside.

5. Repeat steps 1 through 4 with the rest of the top-drilled beads.

Note

Materials listed are for a 20-inch necklace. Add or subtract materials for a longer or shorter necklace.

3. Separate the wires. Pass one wire through one side of the beaded star, and the other wire through the other side of the star (figure 3).

fig. 3

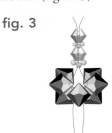

4. String one 4 mm spacer bead, 1 fuchsia 4 mm bicone bead, one 3 mm spacer bead, 1 black 4 mm bicone bead, one 4 mm spacer bead, 1 black 4 mm cube bead, one 4 mm spacer bead, 1 black 4 mm bicone bead, one 3 mm spacer bead, 1 fuchsia 4 mm bicone bead, one 4 mm spacer bead, and 1 fuchsia 8 mm cube bead.

5. Make sure all the beads are snug against the bead stopper. Separate the wires. Use the .010 wire to string 3 size 15° seed beads, 1 size 11° seed bead, one 3 mm spacer bead, 1 black 3 mm bicone bead, one 3 mm spacer bead, 1 size 11° seed bead, and 3 size 15° seed beads. Pass the wire through the fuchsia 8 mm cube bead again and pull tight. This will close one side of the cube. Repeat this 4 more times to completely encase the cube on each side (figure 4). String one 4 mm spacer bead.

fig. 4

6. Repeat steps 2 through 5 five times.

7. Repeat steps 2 and 3.

8. Repeat step 2.

9. Pair the wires together and string 1 crimp tube and half of the clasp. Pass back through the tube, snug the beads, and crimp. Remove the bead stop and repeat at that end of the necklace. Trim any excess wire.

10. Place crimp covers over the crimp tubes.

Stringing the Necklace

1. Pair the beading wires together and place a bead stop or clip about 1½ inches from the wire ends.

2. Keeping the opposite wire ends paired, string 1 fuchsia 4 mm bicone bead, one 3 mm spacer bead, 1 black 4 mm bicone bead, one 4 mm spacer bead, 1 black 4 mm cube bead, one 4 mm spacer bead, 1 black 4 mm bicone bead, one 3 mm spacer bead, 1 fuchsia 4 mm bicone bead, and one 4 mm spacer bead.

Making the Bracelet

Materials for Bracelet

50 black 6 mm top-drilled crystal bicone beads

7 fuchsia 8 mm crystal cube beads

14 black AB2X 4 mm crystal bicone beads

24 black 3 mm crystal bicone beads

12 sterling silver 4 mm Bali-style spacer beads

16 sterling silver 3 mm Bali-style spacer beads

2 sterling silver crimp tubes

1 sterling silver toggle clasp

Beading line, 6-pound test

10 inches of .015 flexible beading wire

Tools

2 big-eye needles

Scissors

Thread burner or lighter

Crimping pliers

Wire cutters

Instructions

1. Make 6 stars as described on page 39. Set aside.

2. Use the wire to string 1 crimp tube and half the clasp. Pass back through the tube, leaving a 1-inch tail. Snug the tube and crimp.

3. String 1 black 3 mm bicone bead, one 3 mm spacer bead, 1 black 4 mm bicone bead, one 3 mm spacer bead, 1 black 4 mm bicone bead, one 3 mm spacer bead, 1 fuchsia 8 mm cube bead, one 3 mm spacer bead, 1 black 4 mm bicone bead, and one 4 mm spacer bead.

4. String 1 star and one 4 mm spacer bead.

5. Repeat steps 3 and 4 five times so you use all 6 stars.

6. Repeat steps 2 and 3 in reverse.

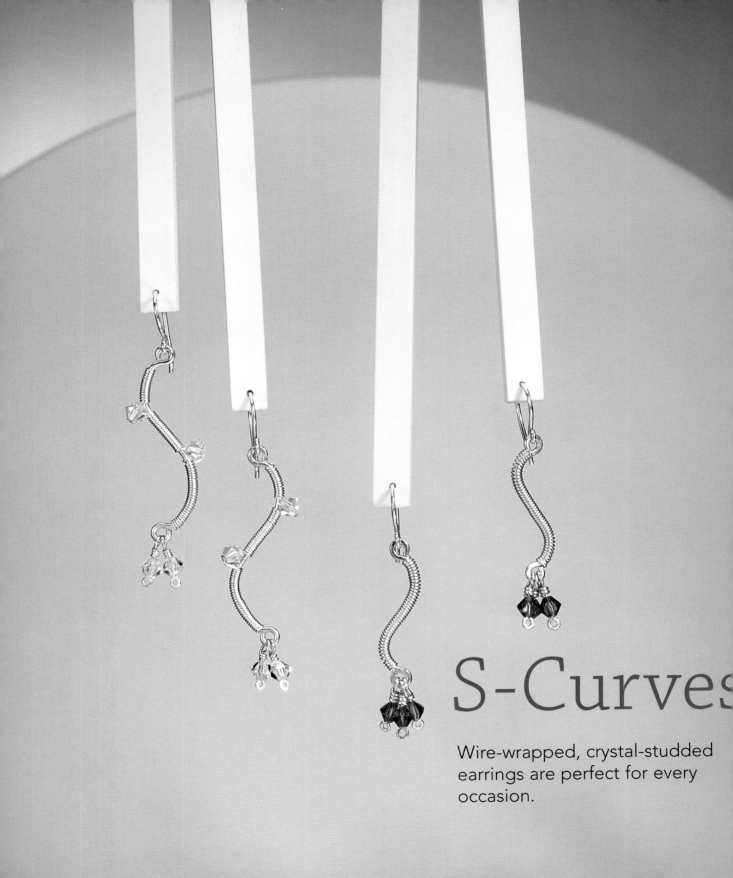

S-Curves

Wire-wrapped, crystal-studded earrings are perfect for every occasion.

Designer: Marie Lee Carter

Materials

6 amethyst 4 mm crystal bicone beads

2 sterling silver French ear wires or lever-back earring findings

4 inches of 18-gauge dead-soft round sterling silver wire

12 inches of 24-gauge dead-soft round sterling silver wire

28 inches of 26-gauge dead-soft round sterling silver wire

Tools

Wire cutters

Ruler

Ball peen hammer

Steel block

Chain-nose pliers

Round-nose pliers

Safety glasses

Barrette file

Long chain-nose pliers

Tumbler (with stainless steel shot, plastic pellets, and gentle liquid soap)

Techniques

Simple loop (page 21)

Wire wrapping (page 21)

Wrapped loop (page 21)

Filing (page 20)

Instructions

1. Cut a 2-inch length of 18-gauge wire. Hammer each end of the wire lightly on the steel block. Use chain-nose pliers to make a 90° bend about ⅜ inch from the flattened end. Use round-nose pliers to grasp the flattened end of the wire about one-third of the way down the jaws and complete a simple loop.

2. Use the hammer to tap the loop two or three times. Make another simple loop on the other end of the wire in the opposite direction. You have just created a wire link.

3. Repeat steps 1 and 2 to make another link to match the first one.

4. Put on the safety glasses. Take the 26-gauge wire and place the end close to an end loop of one of the links. Leave a short tail. Use your fingers to tightly wrap the 26-gauge wire down the body of the link. The result should be a tight, uniform coil. Trim any excess wire and file smooth (figure 1).

fig. 1

5. Hold the links side by side and use your fingers or long chain-nose pliers to gently bend the link loops in opposite directions to create a wave (figure 2). The two links should be as similar as possible.

fig. 2

6. Make a tiny simple loop at the end of the 24-gauge wire. Hammer once. String 1 bead. Begin a wrapped loop, but before closing the loop, attach it to an end loop on one of the links. Complete the wrap. Repeat twice so you have 3 beaded dangles at the end of a link.

7. Repeat step 6 for the second link.

8. Open the loop on one of the earring findings. Attach one of the links and close the finding's loop. Repeat to attach the remaining link to the other earring finding.

9. If you have a tumbler, tumble your new earrings for 45 minutes. Remove and let dry.

Variation

You can easily add beads to the wave portion of this design. Begin the earring as described on page 53, but make the wave before wrapping the wire link. Use a permanent marker to mark the 18-gauge wire ½ inch from the top of the wave, and ½ inch from the bottom. Use the 26-gauge wire to wrap the link as before, but string one 4 mm bead at the first mark (figure 3).

fig. 3

While holding the bead in place, continue to wrap the wave, checking for neatness and uniformity. String another crystal bead at the next mark and finish the wrap, snipping the wire and filing the ends as needed. *Note:* For the earrings to mirror each other, you'll need to make your wraps to the right on one earring and to the left on the other. Finish the earrings as above.

Rock Crystal

On these earrings, rock crystals move and
catch the light beneath wrapped rings.

Rock Crystal

Designer: Eni Oken

Materials

2 rock crystal ½ x ⅝ inch top-drilled briolette or flat pear beads

20 clear AB 2.5 mm crystal bicone beads

2 Bali-style sterling silver 9 mm round bead with braided design

4 sterling silver 10 mm soldered-closed jump rings

2 sterling silver lever-back earring findings

8-inch length of 26-gauge soft round sterling silver wire

5-foot length of 28-gauge soft round sterling silver wire

Tools

Wire cutters

Chain-nose pliers

Round-nose pliers

Techniques

Coiling (page 21)

Wrapped loop (page 21)

Instructions

1. Cut a 2½-foot piece of 28-gauge wire.

2. Stack 2 jump rings. Take the end of the wire and hold it against the jump rings with your left hand, leaving a 1-inch tail. Use your right hand to tightly wrap the jump rings together halfway around the circle. Make the coils tight, but not too close together.

3. When you've reached halfway around the circle, form a tiny wrapped loop with the wire (figure 1).

fig. 1

4. Continue coiling the 2 jump rings together to the top. Form another wrapped loop directly opposite the first one (figure 2). Neatly trim any excess wire. Wrap the very end of the remaining 28-gauge wire around the coiled jump rings, near the top wrapped loop.

fig. 2

5. String 1 bicone bead and pass through the center of the ring and back up, letting the bead sit on the outside of the ring (figure 3). Repeat around 4 more times until you reach the bottom wrapped loop.

fig. 3

6. Pass behind the wrapped loop and secure 5 more bicone beads until you reach the top wrapped loop again. Wrap the wire once around the shank of the top wrapped loop.

7. String 1 silver bead. Bring the wire straight down and wrap it around the shank of the bottom wrapped loop (figure 4). Pass back through the silver bead and wrap the wire around the shank of the upper wrapped loop. Neatly trim any excess wire.

fig. 4

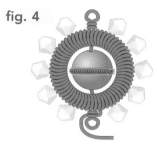

8. Attach the earring finding to the top wrapped loop. Set aside.

9. Cut a 4-inch piece of 26-gauge wire. String 1 briolette, leaving a 1-inch tail. Bring both wire ends together and wrap the short tail around the long one so the wrap sticks straight up from the top of the briolette. Neatly trim the excess short wire close to the wrap.

10. Form a wrapped loop with the long wire, attaching the loop to the bottom wrapped loop of one of the embellished rings. Neatly trim the excess wire close to the wrap.

11. Repeat steps 1 through 10 to make the second earring.

Bling

The design of these wild and wondrous rings allow you to play with various color combinations.

Designer: Wendy Witchner

Materials

10 to 15 assorted color 6 mm crystal bicone beads

15 to 20 assorted color 4 mm crystal bicone beads

10 to 15 sterling silver 3 to 6 mm Bali-style spacer beads

22 sterling silver 1-inch head pins

6 inches of 18-gauge sterling silver round wire

12 inches of 24-gauge sterling silver twisted wire

Tools

Wire cutters

Chain-nose pliers

Round-nose pliers

⅜-inch-diameter dowel

Ring mandrel

Techniques

Simple loop (page 21)

Wrapped loop (page 21)

Coiling wire (page 21)

Instructions

Making the Dangles

1. Use 1 head pin to string an assortment of crystal and silver beads to equal about ⅜ inch. Make a wrapped loop to secure the beads. Repeat with the remaining head pins and beads. Set the dangles aside.

Making the Ring Band

1. Make a simple loop at the end of the 18-gauge wire.

2. Starting at the base of the simple loop, coil the 24-gauge twisted wire tightly around the 18-gauge wire for about 2½ inches (figure 1).

fig. 1

3. Wrap the coiled wire around the ring mandrel (or your finger) so it fits loosely (the crystals will take up some room when you wear the ring). Subtract about ½ inch from that measurement for the ring top. Once you come upon your final figure, adjust the coil wrapping as necessary to finish the ring band. Trim any excess twisted wire and use the chain-nose pliers to tighten the last coil.

Making the Ring Top

1. Use the dowel to form a large loop at the end of the 18-gauge wire (this is the beginning of a wrapped loop). String half of the dangles onto the loop (figure 2).

fig. 2

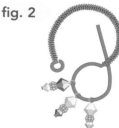

2. Open the simple loop at the other end of the band and attach it to the large loop so the dangles are captured between the coiled portion and the end of the wire. String the other half of the dangles and finish the wrapped loop.

3. Turn the ring top so it is perpendicular to the ring band.

Springy

Show off a variety of beads and charms
with this ultra-fun bracelet.

Designer: Terry Stumpf

Materials

6 blue-gray 5 x 9 mm handmade lampworked beads

6 sterling silver 4 x 8 mm large-holed Bali-style spacer beads

Assortment of 104 clear and gray 3 mm crystal bicone beads

Assortment of 55 clear, gray, and deep blue 4 to 6 mm crystal beads in various shapes

10 sterling silver 4 to 6 mm large-holed beads in various shapes

Assortment of 48 sterling silver charms and/or crystal bead dangles made with 1½-inch head pins

4 gray 4 mm bicone crystals

6 sterling silver 2 x 2 mm crimp tubes

1 sterling silver 12 x 15 mm 2-strand clasp

72 inches of flexible beading wire

12 pieces of ¾-inch sterling silver coiled wire

Tools

Wire cutters

Bead stopper or other clip

Chain-nose pliers

Round-nose pliers (if making dangles)

Crimping pliers

Techniques

Stringing (page 14)

Right-angle weave (page 19)

Wrapped loop (page 21)

Instructions

1. Cut the beading wire in half. Place a bead stopper on one end of each wire, leaving 2- to 3-inch tails.

2. Use 1 wire end to string ¾ inch of assorted crystals. Cross the other wire end through the beads you just strung. Pull the wires tight. String one 3 mm bicone bead, 1 dangle, and one 3 mm bicone bead on each wire end.

3. **fig. 1**

Use 1 wire end to string 1 piece of silver coil and 1 lampworked or silver bead. (The bead will slide over the coil.) Cross the other wire end through the coil. Pull the wires tight. String one 3 mm bicone bead, 1 dangle, and one 3 mm bicone bead on each wire end (figure 1). *Note:* To make a dangle, use 1 head pin to string a ⅜-inch length of beads. Make a wrapped loop to secure the beads.

4. Repeat steps 2 and 3 until you reach the desired length, minus the width of the clasp. If necessary, add additional rows to make a larger bracelet (the one shown is 6½ inches long).

5. String one 3 mm bicone bead, one 4 mm bicone bead, and one 3 mm bicone bead on each wire end. Use a crimp tube to attach each wire to half of the clasp. Remove the bead stopper from the other end of the wire and use crimp tubes to attach the wires to the other half of the clasp (figure 2).

fig. 2

Art Deco

The clean lines of Art Deco architecture inspired this necklace design.

Designer: Betcey Ventrella

Materials

1 light emerald 9 x 27 mm crystal baguette

1 purple iridescent 18 mm crystal rivoli

1 deep blue iridescent 14 mm crystal rivoli

6 deep purple 6 mm crystal rondelle beads

50 deep purple 6 mm crystal round beads

12 light emerald 6 mm crystal round beads

Dark silver metallic size 11° Japanese seed beads

Dark silver metallic size 15° Japanese seed beads

4 sterling silver 2 x 2 mm crimp tubes

1 sterling silver 14 mm toggle clasp

Size D silver or gray beading thread

20 inches of .019 beading wire

Tools

Scissors

Size 11 or 12 beading needles

Chain-nose or crimping pliers

Wire cutters

Techniques

Right-angle weave (page 19)

Tubular peyote stitch (page 18)

Flat even-count peyote stitch (page 17)

Zipping up peyote stitch (page 18)

Instructions

Stitching the Bezels

1. Thread the needle with 4 feet of thread. Use size 11° seed beads to work 1 row of right-angle weave 22 to 23 units long. The row should be long enough so it just barely wraps around the perimeter of the baguette. Don't concern yourself with a perfect fit at this point; you can make adjustments later.

2. Work a second row of right-angle weave off of the first one.

3. Curve the beadwork into a ring so the first and last units of the rows touch. With the thread exiting up from the end bead on the top left (the last unit added), string 1 size 11° seed bead and pass down through the end bead on the top right (the first unit added in the second row). String 1 size 11° seed bead and pass up through the top left bead and the bead first strung in this step. Pass down through the top right bead. Pass through the bead just added, and the bottom left end bead on the first row.

String 1 size 11° seed bead and pass up through the bottom right end bead, through the last bead added in the previous step, the bottom left end bead, and the bead just added (figure 1). Weave through all again to reinforce.

fig. 1

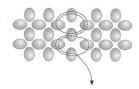

4. String 1 size 15° seed bead and pass through the next bead at the bottom of the right-angle-woven circle (figure 2). Repeat around, working tubular peyote stitch. Pull the thread tight. Step up to the next round. Work a second round of tubular peyote stitch. Weave through all the beads just added once more to reinforce the circle. Weave through the beads and exit from a bead at the top of the circle.

fig. 2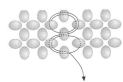

5. Place the baguette, foil side down, into the bowl shape. Work 2 rounds of tubular peyote stitch off the top of the right-angle-woven circle as you did at the bottom of the circle. This creates the baguette's bezel. Reinforce the beadwork once or twice to tighten, if necessary. Secure the thread and trim it close to the beads. Set the bezeled baguette aside.

6. Follow steps 1 through 5 to bezel each rivoli. You'll need to adjust the number of units that encircle the perimeter of your crystals: For the large rivoli, about 20 units will be sufficient, for the small one, about 15.

Connecting the Crystals

1. Start a new thread that exits from a size 11° seed bead at the perimeter of the 14 mm rivoli bezel. String 1 size 11° seed bead and pass through a size 11° seed bead at the perimeter of the 18 mm rivoli bezel. String 1 size 11° seed bead and pass through the seed bead you first exited in this step to connect the two pieces of beadwork (figure 3). Weave through this connection several times until the beads are filled with thread.

fig. 3

2. Weave through the beads to exit from a size 11° seed bead at the perimeter of the 18 mm rivoli bezel, exactly opposite the place you made the connection in step 1. Follow the same directions from step 1, this time making a connection between this bezel and the long side of the baguette bezel. Weave through the connection until the beads are filled with thread. Secure the thread and trim it close to the beads. Set aside.

Adding the Bail

1. Thread the needle with 3 feet of thread. String 9 size 11° seed beads. Work a strip of flat odd-count peyote stitch 9 beads wide by 37 beads long.

2. Sew the last row of the strip to the size 11° seed bead along the back top side of the baguette's bezel, lacing the beads on one piece to the beads on the other so they interlock like a zipper. *Note:* You want the connection to be fairly hidden, so be sure to position it on the back side of the baguette. Curve the strip and sew its first row to the size 15° seed beads at the back top side of the baguette bezel as you did before. Secure the thread and trim close to the beads.

Stringing the Necklace

1. Use the beading wire to string 2 crimp tubes and half of the clasp. Pass back through the tubes, leaving a 1-inch tail. Snug the tube and use chain-nose pliers to squeeze the tube flat, or use crimping pliers to crimp. Trim the tail wire close to the crimp, or cover it with the beads strung in the next step.

2. String a sequence of 1 light emerald 6 mm round bead, 1 size 11° seed bead, 1 rondelle, 1 size 11° seed bead, light emerald 6 mm round bead, 1 size 11° seed bead, and 10 deep purple 6 mm round beads five times.

3. String 1 size 11° seed bead, 1 light emerald 6 mm round bead, 1 size 11° seed bead, 1 rondelle, 1 size 11° seed bead, 1 light emerald 6 mm round bead, the beaded pendant (slide it over the round beads), 2 crimp tubes, and the other half of the clasp. Pass back through the crimp tubes, snug all the beads, and crimp.

Cube crystals and curved wire lines are combined to make these ear adornments.

Designer: Wendy Witchner

Materials for Earring 1

2 fuchsia AB 8 mm crystal cube beads

2 fuchsia 6 mm crystal bicone beads

4 gold-filled 3 mm seamless round beads

2 gold-filled 2 mm seamless round beads

2 gold-filled 2-inch-long paddle head pins

8 inches of 18- or 20-gauge gold-filled wire

Tools

Wire cutters

Round-nose pliers

Chain-nose pliers

⅜-inch-diameter dowel

Chasing hammer

Steel block

Techniques

Simple loop (page 21)

Spiral (see instructions)

Making Earring 1

Instructions

1. Cut the wire into two 4-inch pieces.

2. Make a small loop at the end of one wire.

3. Grasp the loop with the chain-nose pliers so it lies flat within the pliers' jaws.

4. Use your thumb to push the wire along the side of the loop (figure 1). Adjust the pliers and push the wire along the loop again. Repeat until your spiral is 3 wraps wide.

fig. 1

5. Use your fingers to gently pull the wraps apart, creating an airy spiral.

6. Repeat steps 2 through 5 for the other wire, doing your best to duplicate the first spiral's shape.

7. Pair the spirals together so their shapes match. Place the dowel so it sits behind the straight wires, on the opposite side of the direction in which you'd push the next spiral wrap. Use your fingers to push both wires back over the dowel to make the exact same bend (figure 2). These will be the ear wires. Trim the wire ends so the ear wires are even. File the wire ends smooth.

fig. 2

8. Place 1 earring on the steel block so only the spiral sits on the block. Gently hammer the wire until it is slightly flat. Repeat with the second earring. Set aside.

9. Use 1 head pin to string 1 round 3 mm bead, 1 cube bead, 1 round 3 mm bead, 1 bicone bead, and 1 round 2 mm bead. Make a simple loop. Repeat to make a second dangle.

10. Slip 1 dangle onto each earring.

Materials for Earring 2

2 olive AB 8 mm crystal cube beads

2 lavender AB 6 mm crystal bicone beads

4 sterling silver 3 mm seamless round beads

4 sterling silver 2 mm seamless round beads

2 sterling silver 1½-inch-long paddle head pins

9 inches of 20-gauge half-hard sterling silver wire

Tools

Wire cutters

Round-nose pliers

Chain-nose pliers

Chasing hammer

Steel block

½-inch-diameter dowel or pen

Techniques

Simple loop (page 21)

Note

To put on an earring, open the small curled lock. Turn the earring upside down and enter the front of the ear at a slightly curved angle. Make a tight turn in toward the ear, causing the long part of the wire to rotate from the front to the back. The earring should now turn itself right side up.

Making Earring 2

Instructions

1. Cut the wire into two 4½-inch pieces.

2. Starting about 1 inch up the wire, gently hammer the wire flat on the steel block for about ¾ inch.

3. String one 3 mm round bead, 1 cube bead, and one 3 mm round bead onto the 1 inch of wire under the hammered portion. Snug the beads down to the hammered portion and make a simple loop to secure the beads.

4. Starting ½ inch up the wire from the hammered portion, use round-nose pliers to make a U-shaped bend so the two sides of the wire are parallel (figure 1). Use the widest part of the round-nose pliers to make a 90°-angle bend up from the U-shaped bend you previously made (figure 2).

fig. 1

fig. 2

5. Loosely wrap the wire around the dowel for 1 rotation, and then make a thinner rotation to create a gentle, long curl (this will be the earring's lock). Lock the wire around the stem wire and trim any excess wire (figure 3).

fig. 3

6. Use 1 head pin to string one 2 mm round bead, 1 lavender bicone bead, and one 2 mm round bead. Attach the dangle to the earring's simple loop.

7. Repeat steps 2 through 6 to make the second earring, this time making the curl turn the opposite way.

Wire Links

Add as many dangles as you wish to the undulating curves of this freeform wire bracelet.

Designer: Tamara L. Honaman

Materials

4 sapphire AB 10 mm top-drilled (front to back) crystal heart beads

1 crystal AB 20 mm top-drilled (front to back) crystal crescent bead

6 sapphire 8 mm crystal bicone beads

12 crystal AB 8 mm crystal helix beads

3 aqua 16 x 11 mm top-drilled (front to back) crystal baroque pendants

36 sterling silver 2½ mm round beads

36 sterling silver 5 mm daisy spacer beads

4 sterling silver 7 x 10 mm spiral charms with loop

4 silver-plated pinch bails

18 sterling silver 2-inch head pins

12 sterling silver 20-gauge 5.5 mm jump rings

16 sterling silver 8 mm jump rings

30 inches of 14-gauge dead-soft sterling silver wire

Tools

Wire cutters

Round-nose pliers

Chain-nose pliers

Flat-nose pliers

Hammer

Steel block

Techniques

Bending wire (page 21)

Opening loops (page 13)

Wrapped loops (page 21)

Instructions

Making the Links

1. Cut a 3-inch piece of wire. Hammer the last ¼ inch of the wire on the steel block. Repeat for the other end.

2. Grasp one end of the wire with the tip of the round-nose pliers. Rotate your wrist to form a small loop. Repeat on the other end of the wire, so each has a loop facing the same direction (figure 1).

fig. 1

3. Grasp the center of the wire with the largest part of the round-nose pliers' jaws. Bend the wire until the 2 small loops meet, making a U shape (figure 2).

fig. 2

4. Use round-nose pliers to grasp one side of the wire ¼ to ½ inch below the bend you just made. Rotate your wrist so the wire end bends downward. Repeat for the other half of the wire to create an M-shaped link (figure 3).

fig. 3

5. Repeat steps 1 through 4 to create 7 or 8 more links, depending on the length you want for your bracelet. Set aside.

Making the Clasp Eye

1. Cut a 1-inch piece of wire. Hammer the last ¼ to ½ inch of the wire on the steel block. Repeat for the other end.

2. Grasp one end of the wire about three-quarters of the way down the round-nose pliers' jaws. Rotate your wrist away from you until the wire touches itself, creating a large loop.

fig. 4

3. Grasp the other end of the wire with the tip of the round-nose pliers and rotate your wrist away from you until the wire touches itself, creating a small loop that goes in the opposite direction (figure 4). Set aside.

Making the Clasp Hook

1. Cut a 2-inch piece of wire. Hammer the last ¼ to ½ inch of the wire on the steel block. Repeat for the other end.

2. Grasp one end of the wire near the tip of the pliers. Rotate your wrist away from you until the wire touches itself, creating a small loop.

fig. 5

3. Grasp the wire about ½ to ¾ inch above the loop you just created with the largest part of the round-nose pliers' jaws, and bend the wire until it meets the small loop you just made. Grasp the other end of the wire near the tip of the round-nose pliers, and bend the wire in the opposite direction from the last bend until the wire touches itself, creating a small decorative loop. Hammer the hooked portion of this piece, if desired (figure 5). Set aside.

Connecting the Links

1. Lay out the links on the work surface so the first link is positioned like an M, the following one like a W, the next like an M, etc.

2. Open the large jump rings. Connect the first 2 links together using 2 jump rings. Close the jump rings (figure 6). Repeat until all the links are connected.

fig. 6

3. Use 2 jump rings to attach the hook half of the clasp to the small loop on one end of the bracelet. Close the jump rings. Use chain-nose pliers to open the end loop on the other end of the bracelet and attach the eye half of the clasp. Close the loop.

Creating the Dangles

1. String 1 round bead, 1 spacer bead, 1 bicone bead, 1 spacer bead, and 1 round bead onto 1 head pin. Begin a wrapped loop. Before completing the wrap, attach the loop to a bend on the first link of the bracelet. Complete the wrap (figure 7). Trim any excess wire and use chain-nose pliers to tuck in the wire end.

fig. 7

2. String 1 round bead, 1 spacer bead, 1 helix bead, 1 spacer bead, and 1 round bead onto 1 head pin. As in step 1, make a wrapped loop that attaches to the same bend you embellished in step 1.

3. Continue making dangles following steps 1 and 2, and add them along the length of the bracelet as you go. *Note:* You'll want to add all of the dangles in the following steps to the same side of the bracelet so they hang properly when worn.

4. Open a small jump ring and connect it to a baroque bead. Attach it to a link on the bracelet and close the jump ring. Repeat to add the remaining baroque beads so they are evenly spaced down the length of the bracelet.

5. Open a bail and connect it to a heart bead. Close the bail. Open a small jump ring and connect it to the bail's loop. Attach the jump ring to a link on the bracelet and close the jump ring. Repeat to add the remaining heart beads so they are evenly spaced down the length of the bracelet.

6. Open a small jump ring and connect it to a charm. Attach it to a link on the bracelet and close the jump ring. Repeat to add the remaining charms so they are evenly spaced down the length of the bracelet.

7. Open a small jump ring and connect it to the crescent bead. Attach it to a link at the end of the bracelet and close the jump ring.

Wire & Crystal

Add sparkle to your wrist with this exotic cuff made
of wrapped coils and beads around a wire frame.

Designer: Wendy Witchner

Materials

62 assorted color 4 to 6 mm crystal bicone beads

38 sterling silver 3 to 7 mm Bali-style spacer beads and bead caps in various shapes

12 inches of 20-gauge round sterling silver wire

36 inches of 22-gauge twisted sterling silver wire

36 inches of 24-gauge half-hard round sterling silver wire (for wrapping)

16 inches of 14-gauge dead soft round sterling silver wire (for cuff frame)

Liver of sulfur

Tools

Wire cutters

Chain-nose pliers

Round-nose pliers

Planishing hammer

Texturizing hammer

Steel block

Wire brush or tumbler

Bracelet mandrel or frozen juice can

Metal hand file

Techniques

Simple loop (page 21)

Coiling wire (page 21)

Instructions

Making the Coils

1. Tightly coil the twisted wire down the length of the 20-gauge wire. Trim the tails close to the coil.

2. Dip the coiled wire into a liver of sulfur solution, rinse, and brush.

3. Slide the coil off the end of the 20-gauge wire and trim off a 6 mm length. Repeat to make 38 coiled tubes in a variety of lengths, each between 4 and 10 mm long (figure 1). Set the coils aside.

fig. 1

Making the Frame

1. Use your fingers to bend the 14-gauge wire in half, making a soft V shape. The wire sides should be about ¼ inch apart. Use chain-nose pliers to make a 45° bend in each wire about ⅜ inch up from the point so the wire ends are farther apart. Make a simple loop at each end of the wire (figure 2).

fig. 2

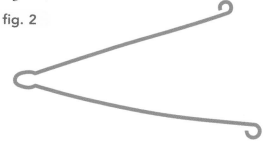

2. Use the hammers and steel block to slightly flatten and texturize the 14-gauge wire.

3. Dip the wire into a liver of sulfur solution, rinse, and brush.

Wrapping the Frame

1. Coil one end of the 24-gauge wire several times to the 14-gauge wire at a spot on the frame just near one of the simple loops.

2. String one 6 mm crystal bead and one 7 mm coil. Draw the beaded wire across toward the other simple loop, gently pull both simple loops together, and make 2 or 3 wraps around the 18-gauge wire from back to front.

3. String one 6 mm crystal bead, one 4 mm bead cap, and one 8 mm coil. Draw the beaded wire across to the other side of the 14-gauge wire and make 2 or 3 wraps from back to front (figure 3).

fig. 3

4. Gently wrap the 14-gauge wire form over the bracelet mandrel or juice can to form a cuff shape.

5. Repeat step 3, stringing beads and coils and wrapping the wire around the opposite side of the cuff. As you move toward the center of the cuff, increase the number of beads and coils used so that by the time you reach the center, the cuff is about 1½ inches wide. Decrease beads as you move toward the other end of the cuff.

6. Once you reach the 45° bend you made when forming the cuff, finish the wire by coiling it several times around one side of the 14-gauge wire.

7. Trim the tail wires at each end of the cuff. Use chain nose pliers to squeeze the beginning and ending coils tight so no wire can abrade the skin. File the wire ends as necessary

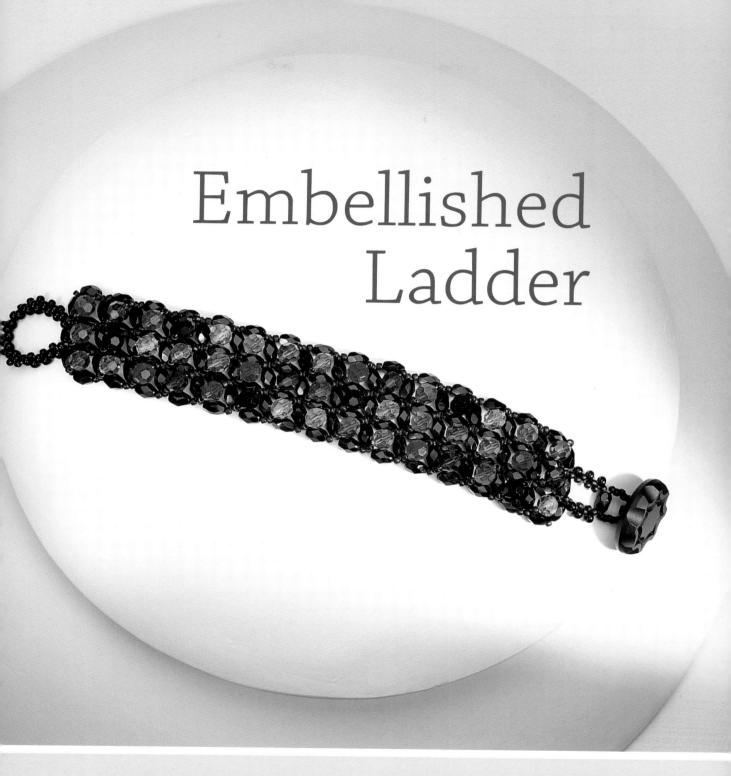

Embellished Ladder

This gorgeous bracelet is made with a traditional beading stitch and shows off crystals in a dramatic way.

Designer: Bonnie Clewans

Materials

45 multicolored 6 mm crystal round beads

108 black 4 x 6 mm fire-polished glass oval beads

2 grams of size 8° Czech or Japanese seed beads

5 grams of size 11° Japanese seed beads

1 glass or metal 18 mm button with shank

.008 beading line

Tools

Scissors

2 size 10 beading needles

Techniques

Right-angle weave (page 19)

Peyote stitch (page 17)

Note

Materials are for a 6½-inch bracelet. Add or subtract beads as necessary to make a larger or smaller bracelet.

Instructions

1. Cut a 6-foot length of line and thread a needle on each end. String the button onto one needle and slide it to the center of the thread.

2. String 5 size 11° beads onto each needle and slide them t the button. String 1 oval bead onto the right needle and cross through the oval bead with the left needle. Tighten the beads.

3. **fig. 1**

String 7 size 11° beads onto each needle, string 1 oval bead onto the right needle, and cross through the oval bead with the left needle (figure 1).

4. **fig. 2**

Set the left needle aside. Pass the right needle back through the last size 11° see bead strung on the right side in the previous step. String 1 size 8° seed bead, skip 1 size 11° seed bead, and pass back through the next size 11° bead. Continu working peyote stitch up the strand. Pa through the first oval bead, the first 5 seed beads added on the left, the button the first 5 seed beads added on the right, and through the first oval bead. Use size 8° seed beads to work peyote stitch down the left side of the strand and exit fro the second oval bead (figure 2)

5. **fig. 3**

String 1 size 11° bead, 1 oval bead, and size 11° bead onto each needle. String 1 oval bead onto the right needle and cros through it with the left needle to make a right-angle weave (figure 3). Repeat 14 more times to make 15 right-angle-wove units in all. Test the bracelet for fit, considering that the loop portion of the clas will add another inch. If necessary, add subtract units.

6. Make the loop for the button/loop clasp. String 28 size 8° beads onto the right needle and pass through the final oval bead added to the last right-angle weave unit. Slide the loop over the button to test for fit. The loop should slide easily over the button, but it shouldn't be too loose. Add or subtract size 8° beads as necessary.

7. Continue to use the right needle. Pass through the first size 8° bead strung in step 6. String 1 size 11° bead, skip the next size 8° bead of the loop, and pass through the next. Continue working peyote stitch around the loop.

8. Use both the right and the left needles to weave through the units to reinforce the beadwork. Secure the thread and trim close to the beads.

> ## Note
>
> If you wish to make the single-stranded version of this bracelet, stop here, work step 13, secure the thread, and trim.

9. **fig. 4**

Start a new single thread at the button end of the bracelet that exits from the second seed bead added to the first unit of the first row. String 1 oval bead, 1 size 11° bead, 1 oval bead, 1 size 11° bead, and 1 oval bead. Pass down through the first seed bead, first oval bead, and second seed bead of the first unit of the first row, and the first seed bead, first oval bead, and second seed bead of the second unit of the first row (figure 4).

10. **fig. 5**

String 1 oval bead, 1 size 11° bead, 1 oval bead, and 1 size 11° bead. Pass through the bottom oval of the previous unit in this row and down through the side beads (seed bead, oval bead, and seed bead) of the adjacent unit and the next unit of the first row (figure 5).

11. Repeat step 10 down the row to add 15 right-angle-woven units in all. Weave through the units of this second row once to reinforce the beadwork. Secure the thread and trim close to the beads.

12. Repeat steps 9 through 11 to add 15 right-angle-woven units down the other side of the bracelet, creating a third row.

13. **fig. 6**

Start a new single thread that exits from the first oval added to the first unit of the first row. String 1 size 11° bead, 1 round bead, and 1 size 11° bead. Cross the strand to the opposite corner of the unit and pass through the oval at the unit's bottom (figure 6). Repeat down the row to add 15 round beads in all.

14. Repeat step 13 down the second row, making sure that the beads are all crossing at the same angle as those in the first row.

15. Repeat step 13 down the third row, again making sure that the beads all lay in the same direction. Secure the thread and trim close to the beads.

Squared Cubes

The designer combined a square stitch with cube-shaped beads, resulting in this lovely pendant.

Designer: Tina Koyama

Materials

23 sapphire AB 5 mm crystal cube beads

3 lavender AB 5 mm crystal cube beads

12 lavender AB 4 mm crystal bicone beads

24 sapphire AB 3 mm crystal bicone beads

Blue size 11° seed beads

2 sterling silver crimp beads

Sterling silver toggle clasp

Braided 6-pound-test beading line

Nylon-coated beading wire (desired length plus 2 inches)

Tools

Scissors

Size 10 beading needle

Crimping pliers

Wire cutters

Techniques

Tension bead (page 15)

Square stitch (page 19)

Securing thread (page 15)

Trimming thread (page 15)

Stringing (page 14)

Crimping (page 14)

Instructions

1. Thread the needle with 3 feet of thread. Use a seed bead to make a tension bead, leaving a 6-inch tail. String 1 sapphire cube bead, 1 lavender cube bead, and 4 sapphire cube beads, forming row 1 (the top of the pendant).

2. String 1 sapphire cube bead, pass through the last bead strung in row 1, and pull snug so the 2 cube beads sit side by side. Pass through the bead you just added again to make your first square stitch. Continue across the row, square-stitching 1 sapphire cube bead to each cube bead in the previous row.

3. Pass through all the beads in rows 1 and 2 again to reinforce the work.

4. Work row 3 by using sapphire cube beads for 4 stitches, and 1 lavender cube bead in the fifth stitch. Make an end-of-row decrease by stopping the row short by 1 cube bead (figure 1).

fig. 1

5. In each succeeding row, decrease by 1 bead until you have 2 beads in row 6. Follow figure 2 for bead color placement.

fig. 2

6. Reinforce the pendant by weaving through each row until the pendant is stable and firm, leaving enough space in the beads of row 1 for the beading wire to pass through. Remove the tension bead. Secure the thread and trim it close to the work.

7. String row 1 of the pendant onto the beading wire and slide it to the center.

8. String 5 seed beads on one end of the wire. String a sequence of one 3 mm bicone bead, 1 seed bead, one 4 mm bicone bead, 1 seed bead, one 3 mm bicone bead, and 10 seed beads 6 times. String seed beads until you reach the desired length of one half of the necklace.

9. String 1 crimp tube, 1 seed bead, and half of the clasp. Pass back through the seed bead and crimp tube just strung. Crimp the tube.

10. Repeat steps 8 and 9 to complete the other side of the necklace.

Fibula

The design of this brooch is inspired by the work of modernist jewelers of the mid-20th century.

Designer: Rachel Dow

Materials

1 amethyst 8 mm x 18 mm top-drilled polygon drop crystal bead

2 amethyst 13 mm x 6.5 mm top-drilled briolette crystal beads

10 inches of 14-gauge round, dead-soft sterling silver wire

6 inches of 22-gauge round, dead-soft sterling silver wire

Tools

Paper and pencil

Wire cutters

Chain-nose pliers

Round-nose pliers

Hammer (chasing and ball peen)

Small anvil

Center punch

Flex shaft with size 70 drill bit

Wooden surface

Metal hand file

P320 emery paper

Techniques

Simple loop (page 21)

Drilling (see instructions)

Filing and sanding (page 20)

Instructions

1. Draw a freeform sketch of the fibula that includes three bottom curves. Include the bead placement on your sketch. You will refer to the sketch when you start bending the wire. Make the drawing so the fibula is about 1¾ inches wide and 1¼ inches high.

2. Use round-nose pliers to make a large simple loop at one end of the 14-gauge wire. This loop will be used for the pin stem catch.

3. Refer to your freeform sketch as you bend the wire with the round-nose pliers. Leave 2 inches of straight wire to make a pin stem (figure 1). Trim any excess wire.

fig. 1

4. Place the fibula on the anvil and use the ball peen hammer to flatten the bottom and top curves, creating a dimpled texture.

5. Mark your fibula where you'd like your holes to be drilled on the three bottom curves. Place the fibula on the anvil and use the center punch and the flat end of the chasing hammer to make a small dimple in the wire. This small mark will help your drill stay steady. Place the fibula on a wooden surface and use the flex shaft to drill the holes.

6. Use the emery paper to lightly sand the surface of the fibula, removing any scratches and creating a brushed finish.

7. Use the chain-nose pliers to open the simple loop created in step 2, making a C shape that's perpendicular to the main design. Make sure the hook is at a right angle to the fibula

8. File the end of the wire for your pin stem. Make the point a gradual taper. The pin point should be about ¾ inch long. Use the emery paper to sand the tip of the stem, removing any scratches.

9. To bend the pin stem, use the round-nose pliers to bend the pin stem back toward the hook. The pin stem should be longer than the bent hook by about ½ inch. Set the fibula aside.

10. fig. 2   Cut a 2-inch piece of 22-gauge wire. String 1 small briolette bead, leaving a 1-inch tail. Bend the wire ends so they meet over the top of the bead. Use the chain-nose pliers to bend each wire so they point straight up over the top of the bead (figure 2).

11. fig. 3   Wrap one wire around the other one to make 2 coils, from the top of the bead up. Neatly trim any excess wire from the part you've been working. Use the chain-nose pliers to gently squeeze the tail close to the stem (figure 3).

12. fig. 4   Make a wrapped loop with the remaining wire, attaching to the first drill hole, and wrapping until you meet the wraps made in the previous step (figure 4). Neatly trim any excess wire.

13. Repeat steps 10 to 12 to add the large polygon bead to the fibula's center hole, and the second briolette bead to the remaining hole.

Linked Squares

Dazzling crystal squares connected by peyote links compose this highly fashionable bracelet.

Designer: Tamara Honaman

Materials

7 dark green 20 mm crystal diamond cutouts

2 olive 4 mm crystal bicone beads

2 metallic green ½-inch bugle beads

12 grams of metallic green cylinder beads

2 moss green size 15° seed beads

Brown beading thread

Beeswax

Tools

Scissors

Size 10 beading needle

Big-eye needle

Chain-nose pliers

Techniques

Peyote stitch (page17)

Picot edging (at right)

Zipping up peyote stitch (at right)

Ladder stitch (page 17)

Instructions

Linking the Squares

1. Use 2 yards of thread to string 6 cylinder beads, leaving a 12-inch tail.

2. Work peyote stitch until you have a strip 6 beads wide by 60 rows long.

3. Exit from a side bead of the strip. String 3 cylinder beads and pass down through the next side bead and up through the following one (figure 1). Continue making this edging until you have 6 picots. Skip 4 side beads, and make 6 more picots. Repeat to make a mirror edging on the other side of the strip.

fig. 1

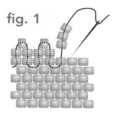

4. Place 2 diamonds back to back and pass the strip through the center.

5. Match the first and last rows of the strip so they interlock like a zipper. Weave back and forth through the beads, lacing the rows together (figure 2). Weave through all the beads again to reinforce. Secure the thread and trim close to the beads.

fig. 2

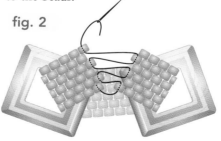

6. Open the diamonds so they sit right side up. Position the areas without picots so they rest on the insides of the diamonds.

7. Repeat steps 1 to 3 to make another strip. Place 1 diamond back to back on the last diamond you added. Pass the new strip through the new and last diamonds and zip the strip together as before.

8. Repeat step 7 until you've linked enough diamonds together to reach your desired length, minus ¾ inch for the toggle. Set aside.

Making the Clasp

1. Use a new thread and cylinder beads to work a strip of peyote stitch 6 beads wide by 28 rows long. Wrap the strip around one side of a diamond at one end of the bracelet, and zip the edges as in step 5.

2. **fig. 3**

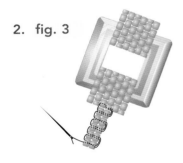

Weave through the beads to exit from the fourth bead across in one row of the strip. (Your needle should be pointing toward the edge.) String 2 cylinder beads and pass through the third bead across the strip, then back out the fourth. Work a strip of ladder stitch off of these beads that is 2 beads high and 5 beads long (figure 3). Don't cut the thread. Set the bracelet aside.

3. Use a new thread and cylinder beads to work a strip of peyote stitch 18 beads wide and 12 rows long.

4. Place the 2 bugle beads on the strip, end to end, spanning the width of the strip. Zip the strip together to make a tube, capturing the bugle beads inside. The bugle beads will act as stabilizers to give the tube strength. Weave through the beads to exit from a bead at one end of the tube. Remove the beading needle and thread on the big-eye needle. Pass through to the other side of the tube.

5. String 1 bicone bead and 1 size 15° seed bead. Pass back through the crystal and out the other end of the tube. Pull the thread tightly so the crystal sits in the tube. Repeat to add a bicone bead to the other end of the tube (figure 4).

fig. 4

Pass back and forth through the bicone beads and seed beads to reinforce. Carefully weave into the beads of the tube, and then reinforce the beadwork. Secure the working and tail threads and trim close to the beads.

6. Thread the beading needle on the working thread at the end of the ladder-stitched strip. Weave through two center beads of one row in the toggle bar and pass through the last two beads added in the ladder-stitched strip. Repeat the thread path as needed to make a secure connection. Weave down the strip to reinforce and back into the peyote-stitched strip attached to the diamond. Secure the thread and trim close to the beads.

Designer's Tip

You will most likely fill the cylinder beads up with thread in this project. In order to avoid bead breakage, be very careful as you pull the thread through the beads, using chain-nose pliers to do the job, if needed.

Pearlescent

Luscious freshwater pearls are combined with the hard-edged perfection of crystals to create this stunning necklace.

Materials

28 pale green 6 x 17 mm stick freshwater pearl beads

7 pale green 6 to 7 mm round freshwater pearl beads

10 light green 6 mm crystal round pearl beads

10 olive AB 6 mm crystal round beads

7 dark amethyst AB 6 mm crystal round beads

7 light amethyst AB 6 mm crystal round beads

5 silver AB 6 mm crystal cube beads

12 amethyst 4 mm bicone beads

16 olive 4 mm bicone beads

Purple iris size 8° seed beads

Opaque green size 11° seed beads

Purple iris size 11° seed beads

12 light matte-green 7 mm pressed-glass flower cap beads

7 amethyst 8 mm pressed-glass flower beads

5 amethyst 14 mm pressed-glass flower beads

5 matte-olive 8 mm round beads

42 sterling silver 24-gauge 2-inch head pins

2 gold-filled 2 x 2 mm crimp beads

14 mm gold box clasp set with peridot inset

22 inches of gold .019 flexible beading wire

Instructions

1. Use 1 head pin to stack an arrangement of beads as outlined below. Secure the beads with a wrapped loop. You will make 30 bead dangles in all.

 Dangle A: String 1 crystal pearl bead and 3 purple size 11° seed beads. (Make 6.)

 Dangle B: String 1 olive 4 mm crystal bicone bead, 1 green size 11° seed bead, and 1 small amethyst flower bead. (Make 6.)

 Dangle C: String 1 amethyst 4 mm crystal bicone bead, 1 matte-green flower bead, and 3 purple size 11° seed beads. (Make 10.)

 Dangle D: String 1 purple size 11° seed bead, 1 purple size 8° seed bead, 1 olive 8 mm round bead, and 1 purple size 11° seed bead. (Make 5.)

 Dangle E: String 1 purple size 11° seed bead and 1 pale green 7 mm freshwater pearl bead. (Make 5).

 Dangle F: String 1 dark amethyst 6 mm crystal round bead and 3 green size 11° seed beads. (Make 5.)

 Dangle G: String 1 olive 6 mm crystal round bead, 2 green size 11° seed beads, and 1 large amethyst flower bead. (Make 5.)

2. Use the beading wire to string 1 crimp bead, 1 crystal pearl bead, and half of the clasp. Pass back through the crystal pearl bead and the crimp bead, leaving a 1-inch tail. Snug the beads and crimp the crimp bead. Cut the tail wire close to the crimp.

3. String 1 amethyst 4 mm bicone bead, 1 green flower bead from inside-to-outside, 1 small amethyst flower bead from inside-to-outside, 1 crystal pearl bead, 1 light amethyst 6 mm bead, 1 stick pearl bead, 1 round pearl bead, 1 purple size 11° seed bead, 2 stick pearl beads, and 1 dark amethyst 6 mm bead.

4. String 1 stick pearl bead, 1 dangle A, 1 dangle B, 1 stick pearl bead, 1 olive 4 mm bicone bead, 1 dangle C, 1 dangle D, 1 light amethyst 6 mm bead, 1 stick pearl bead, 1 dangle E, 1 dangle F, 1 stick pearl bead, 1 olive 4 mm bicone bead, 1 dangle G, 1 dangle C, and 1 cube. Repeat the sequence 4 more times to make 5 repetitions in all.

. String 1 stick pearl bead, 1 dangle A, and 1 dangle B.

. Repeat step 3 in reverse to string the other side of the necklace.

. String 1 crimp bead, 1 crystal pearl bead, and the other half of the clasp. Pass back through the crystal pearl bead and the crimp bead.

Before crimping, hold the necklace up by the unfinished end. Use your fingers to adjust the beads so they sit nicely together, with little or no gaps between each. To make sure the beads are settled, but not strung too tightly, let the piece coil gently into your hands. If this movement isn't fluid, loosen the spacing a bit to reduce the tension on the beading wire. This will not only help with drape and comfort while you're wearing it, but will also help avoid wire abrasion and breakage.

. Crimp the crimp bead. Trim any excess wire close to the crimp.

Tools

Chain-nose pliers

Round-nose pliers

Crimping pliers

Wire cutters

Techniques

Wrapped loop (page 21)

Stringing (page 14)

Crimping (page 14)

Blossoming

In this necklace, a delicate band of flowers contrasts with brilliant crystals and sterling silver chain.

Materials

3 peridot 10 mm crystal heart beads, horizontally drilled from back to front

4 peridot 6 mm crystal bicone beads

11 rose 4 mm crystal round beads

3 double-drilled 15 mm acrylic flower slider beads with rose crystal insets

3 sterling silver 2-inch head pins

2 sterling silver 2-inch eye pins

2 sterling silver 20-gauge 2 mm jump rings

11 mm sterling silver lobster clasp and ring

10 inches of 20-gauge sterling silver wire

13½ inches of 10 mm sterling silver 2.3 mm link chain

Tools

Wire cutters

Chain-nose pliers

Round-nose pliers

Techniques

Stringing (page 14)

Simple loops (page 21)

Opening loops (page 13)

Instructions

1. Cut one 1-inch, one 1½-inch, and two 5½-inch pieces of chain. Set aside.

2. Make a tiny simple loop on one end of the sterling silver wire. Use the wire to string 1 flower through the bottom 2 holes, 2 rose 4 mm beads, an end link on the 1-inch piece of chain, 1 flower through the bottom 2 holes, an end link on the 1½-inch piece of chain, 2 rose 4 mm beads, and 1 flower through the bottom 2 holes. Work with the natural curve of the wire and use your thumbs as support as you bend the two ends of the wire upward to make an inverted arch. Cut the wire about ³⁄₁₆ inch from the last bead strung.

3. Make a normal-size simple loop at one end of the remaining sterling silver wire. Pass through the top 2 holes of the first flower strung in the previous step. String 1 rose 4 mm bead. Pass through the top 2 holes of the second flower strung in the previous step. Pass through the top 2 holes of the third flower strung in the previous step.

4. Use your fingers to adjust the degree of arch in the piece. Once you are satisfied, finish the top wire with a simple loop and the bottom wire with a tiny loop (figure 1). Set this focal piece aside.

fig. 1

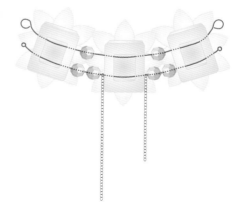

5. Use 1 eye pin to string 1 peridot 6 mm bead, 1 rose 4 mm bead, and 1 peridot 6 mm bead. Make a simple loop to secure the beads in place. Repeat to make a second 3-bead link and set aside.

6. Open the loop on one end of a 3-bead link and connect it to an end link of one of the 5½-inch pieces of chain. Close the loop. Open the loop on the other side of the same 3-bead link and connect it to the top loop at one end of the focal piece. Repeat to add a link and chain to the other side of the focal piece.

7. Use a jump ring to connect one half of the clasp to the end of one of the 5½-inch pieces of chain. Repeat for the opposite chain and clasp end.

8. Use 1 head pin to string 1 heart bead. Carefully manipulate the pin so it bends up and over the top of the heart, right at the V. Grasp the wire with chain-nose pliers and make a 90° bend just above the heart. String 1 rose 4 mm bead and secure the beads with a simple loop. To make the heart hang face forward, position the loop so it's perpendicular to the front of the heart (figure 2). Repeat to make two more heart dangles.

fig. 2

9. Open the loop on one of the heart dangles and connect it to the jump ring next to the clasp. Attach the remaining two hearts to the ends of the chains hanging from the focal piece.

Brilliant Fuchsia

This beaded wheel pin is created with a circular brick stitch and accented with vivid fuchsia crystals.

Designer: Tina Koyama

Materials

1 fuchsia AB 10 mm crystal round bead

12 fuchsia AB 6 mm crystal bicone beads

12 fuchsia AB 4 mm crystal bicone beads

Magenta-lined size 8° hex beads

Gold-lined size 8° twisted hex beads

Gold iris shiny size 11° twisted hex beads

Gold iris matte size 11° hex beads

Fuchsia size 11° seed beads

Gold iris cylinder beads

Gold transparent size 14° seed beads

1½-inch silver pin back

Size D (.008) braided beading line

Tools

Scissors

Size 10 beading needle

Techniques

Circular brick stitch (page 16)

Brick stitch increase (page 16)

Fringe (page 15)

Instructions

Preparing the Center

1. Thread the needle with 3 feet of thread. String the 10 mm round bead, leaving a 6-inch tail.

2. Pass through the 10 mm bead again so that the thread wraps around one side of the bead. Pass through the bead again so that the thread wraps around the other side of the bead.

3. fig. 1 Repeat step 2 so that 2 strands of thread are on each side of the bead. Pull the thread tight and tie an overhand knot around the threads to stabilize the stitching (figure 1).

Stitching the Rounds

1. Work circular brick stitch around the 10 mm bead.

fig. 2 **Round 1:** Working with cylinder beads and using the double threads from step 3 as your base, stitch 19 beads around. Make the step up to the next round by passing down through the first bead added, and up through the second bead in the round (figure 2).

Round 2: Using gold iris matte size 11° hex beads, make an increase about halfway around the circumference of the work to add 20 beads in all.

Round 3: Using gold iris shiny size 11° twisted hex beads, add 26 beads in all.

Round 4: Using gold-lined size 8° twisted hex beads, add 26 beads in all.

Round 5: Using fuchsia size 11° seed beads, add 38 beads in all.

Round 6: Using fuchsia size 11° seed beads, add 43 beads in all.

Note

Although it's important to place an accurate number of beads in each round to maintain the pattern, where you place each increase is not as important. Just be sure to spread the increases around the circumference fairly evenly. For example, if you know you will need to increase the successive round by 4 beads, you can place the first increase approximately one-fourth of the way around, the second increase about halfway around, and so on. With experience, you will start to know when an increase is needed because a loop of thread your needle goes under will seem longer than the others, which means it's a good place to put an increase. When you are about halfway around the circumference, you may want to count the number of beads you have placed and the number of loops of thread remaining to plan the placement of the remaining increases needed.

Round 7: Using gold iris matte size 11° hex beads, add 48 beads in all.

Round 8: Using magenta-lined size 8° hex beads, add 48 beads in all.

. Secure the thread, but don't trim. Exit from the top of any bead in round 8.

Embellishing the Pin

1. String 1 size 11° seed bead, one 4 mm bicone bead, and 3 size 14° beads. Pass back through the bicone bead. String 1 size 11° seed bead and pass down through the adjacent hex bead in round 8. Pass up through the next hex bead. String 1 size 11° seed bead, one 6 mm bicone bead, and 3 size 14° beads. Pass back through the bicone bead. String 1 size 11° seed bead and pass down through the adjacent hex bead in round 8. Pass up through the next hex bead (figure 3).

fig. 3

2. Repeat step 1 to add alternating fringe legs around the circumference of the circle. Secure the thread, but don't trim.

Attaching the Pin

1. Open the pin back and place it on the back of the beadwork, just above the center 10 mm bead. Weave through the beads until you reach a spot on the beadwork near a hole on the left end of the pin back.

2. String 3 size 11° seed beads, pass through the pinback hole, and continue passing through to the right side of the beadwork. Weave through 1 or 2 beads and pass back through the beadwork to the back side, being careful not to expose the thread. String 3 size 11° seed beads, and pass through the same pin back hole, this time from the other side of the pin back.

3. Repeat step 2 for each of the pin back's holes. If the pin back feels wobbly, reinforce the stitching by weaving through all again.

Caged Crystals

To make these earrings, assemble five bead strands, caging a central crystal.

Designer: Anna Elizabeth Draeger

Materials

2 deep blue AB 6 mm crystal bicone beads

16 emerald AB 5 mm crystal bicone beads

4 dark amethyst 4 mm crystal bicone beads

8 dark amethyst 4 mm crystal polygon beads

20 silver size 11° seed beads or 2 mm sterling silver beads

4 sterling silver 2 x 2 mm crimp beads

4 sterling silver crimp bead covers

2 sterling silver earring findings

60 inches of .010 beading wire

Tools

Wire cutters

Crimping pliers

Chain-nose pliers

Techniques

Stringing (page 14)

Crimping (page 14)

Instructions

1. Cut the beading wire into five 12-inch pieces.

2. Gather the ends of all 5 beading wires and string 1 crimp bead. Slide it about 2 inches down over the wires. Pull one of the wires out another inch or so. Pass this wire back through the crimp bead and crimp (figure 1). Trim the remaining 4 strands right next to the crimp bead, taking care not to cut the loop of beading wire.

fig. 1

3. String one 4 mm bicone bead and 1 seed bead over all 5 strands. Separate the strands. On the first strand, string 1 seed bead, one 6 mm bicone bead, and 1 seed bead (figure 2).

fig. 2

4. On each of the remaining 4 strands, string one 5 mm bicone bead, 1 seed bead, one 4 mm polygon bead, 1 seed bead, and one 5 mm bicone bead.

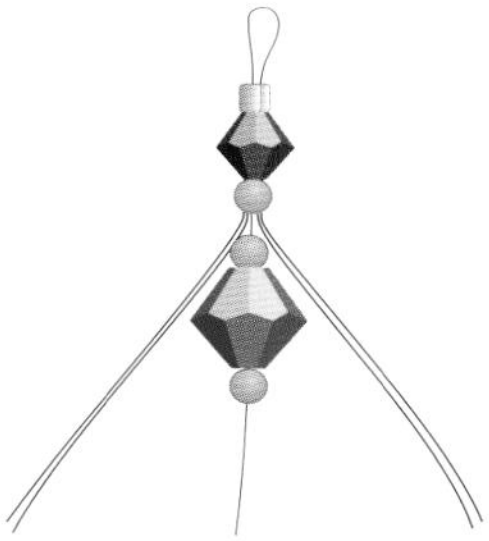

5. Gather the ends of all 5 strands and string 1 seed bead, one 4 mm bicone bead, and 1 crimp bead (figure 3). Arrange the strand with the sole 6 mm bead so it sits in the center of the other 4 strands, and snug all the beads. Adjust the tension so no beading wire shows. Crimp the crimp bead and trim the wires right next to the crimp.

fig. 3

6. Use chain-nose pliers to close the crimp bead covers over the crimp beads.

7. Attach the earring finding to the loop of beading wire.

8. Repeat steps 1 through 7 to make a second earring.

Denim Lace

This necklace pairs a netted configuration of cylinder beads with the understated elegance of denim blue crystals.

Designer: Anna Elizabeth Draeger

Materials

66 denim blue AB2X 4 mm crystal bicone beads

5 grams of Japanese cylinder beads (A)

2 grams of Japanese cylinder beads (B)

1 sterling silver 4 mm soldered jump ring

2 sterling silver 2 x 2 mm crimp tubes

1 sterling silver 9 mm lobster clasp

24 inches of .012 flexible beading wire

Nylon beading thread

Tape

Tools

Wire cutters

Crimping pliers

Scissors

Size 12 beading needle

Techniques

Stringing (page 14)

Crimping (page 14)

Netting (page 17)

Overhand knot (page 14)

Instructions

Stringing the Base

1. String 1 crimp tube and the clasp onto the beading wire, leaving a 1-inch tail. Pass back through the tube and crimp. Trim the tail next to the crimp tube.

2. String 25 A beads, 1 B bead, 1 crystal bead, 1 B bead, 30 A beads, 1 B bead, 1 crystal bead, 1 B bead, and 35 A beads.

3. String a sequence of 1 crystal bead and 1 A bead 32 times.

4. Repeat step 2 in reverse. Secure the end with tape.

Stitching the Base

1. Thread the needle with 3 yards of thread, leaving a long tail to be woven in later. Weave through the beads to exit from the last cylinder bead added before the series of 32 crystals.

2. String 1 B bead, 5 A beads, and 1 B bead. Pass through the cylinder bead between the next 2 crystal beads and back through all the beads just strung (figure 1). Pass through the first 4 beads strung in this step.

fig. 1

3. String 2 A beads, 1 B bead, and 3 A beads. Pass back through the last B bead added to form a picot.

4. String 5 A beads and pass through the last B bead added in step 2 and through the next crystal bead (figure 2).

fig. 2

5. String 1 B bead and 2 A beads. Pass back through the third A bead added in the previous step. String 2 A beads, 1 B bead, 1 crystal bead, and 1 B bead. Skip the last B bead added and pass back through the crystal bead just added. String 5 A beads. Pass through the first B bead added in this step and the next crystal bead (figure 3).

fig. 3

6. Repeat step 5 once.

7. Repeat step 5 four times, but add 2 crystal beads instead of 1.

8. Repeat step 5 once, but add 3 crystal beads instead of 1.

9. Repeat step 5 once, but add 4 crystal beads instead of 1. This is the center of the necklace.

10. Repeat steps 1 through 8 in reverse to mirror the first side of the netting. Secure the thread and trim close to the beads.

Finishing the Necklace

1. Adjust the strung and netted portions of the necklace so they are even along the beading wire. Remove the tape.

2. String 1 crimp tube and the jump ring onto the beading wire. Snug the beads and crimp the tube.

Fringe

Topaz and gold crystals add sparkle and flair to this handsome
bracelet, which is finished with a dramatic clasp.

Designer: Tamara Honaman

Materials

2 smoky topaz 6 mm crystal bicone beads

About 50 light smoke 3 mm crystal bicone beads

About 50 light smoky topaz 3 mm crystal bicone beads

About 50 smoky topaz AB 3 mm crystal bicone beads

About 500 metallic gold 3 mm crystal bicone beads

About 50 clear rainbow dark amber Japanese fringe beads

Size 8° iris bronze seed beads

Size 11° iris metallic yellow hank seed beads

Size 15° gold luster rainbow gold seed beads

6 gold-filled 3 mm round beads

4 gold-filled 2 mm crimp beads

4 gold-plated 3 mm crimp covers

16 x 24 mm magnetic clasp with smoky topaz domed crystal inset

18 inches of gold-plated .019 flexible beading wire

Beading thread to coordinate with beads

Instructions

Creating the Core

1. Measure your wrist. Add 2 inches and subtract the length of the clasp. Cut 2 pieces of beading wire to that length.

2. String 1 crimp tube and 1 gold 3 mm round bead. Pass through one hole on the bottom half of the clasp and back through the beads just strung. Snug the wire and crimp the tube. Repeat using the other length of wire, connecting to the other hole on the same side of the clasp.

3. Bring both wires together and string 1 gold 3 mm round bead, 1 smoky topaz 6 mm bicone bead, and 1 gold 3 mm round bead (figure 1).

fig. 1

4. Continue stringing over both wires, adding about 6 inches of size 8° seed beads (about 60 beads), 1 gold 3 mm round bead, 1 smoky topaz 6 mm bicone bead, and 1 gold 3 mm bead. Test for fit, keeping in mind that you will still be adding the length of 1 gold 3 mm bead and 1 crimp tube. If necessary, adjust the length by adding or subtracting size 8° beads.

5. Separate the wire ends. String 1 crimp tube and 1 gold 3 mm round bead on one wire. Pass through the corresponding hole on the other half of the clasp and back through the beads just strung. Repeat the stringing sequence for the other wire. Make sure you leave some slack in the strand (about the width of a size 8° bead) so you'll be able to get your needle through the beads when you're stitching the fringe (in the next section).

Adding the Fringe

1. Thread the needle with about 3 yards of thread. Pass it through the first size 8° bead in the core, leaving a 4-inch tail.

2. String 4 size 11° beads and one size 15° bead. Skip the size 15° bead and pass back through the size 11° beads. Exit from the next size 8° bead on the core.

3. String 6 size 11° beads, 1 bicone 3 mm bead (any color), and 1 size 15° bead. Pass back through the crystal and the size 11° beads. Exit through the next size 8° bead on the core (figure 2).

fig. 2

4. Repeat step 3, this time replacing the 3 mm bead with a fringe bead.

5. Continue adding fringe down the length of the core. Vary the end bead on each fringe leg to create texture and interest. When you reach the end of the core, exit from the last size 8° bead.

6. Reverse your thread direction by adding a fringe leg out of the last size 8° bead. Work back down the core, adding a row of fringe that matches the first one. Add new thread as needed. When you get to the end, reverse your thread direction and make another row of fringe. Add as many rows as desired until you're pleased with the bracelet's fullness.

Tools

Flexible measuring tape

Wire cutters

Crimping pliers

Size 12 beading needle

Scissors

Chain-nose pliers

Techniques

Stringing (page 14)

Fringe (page 15)

Adding thread (page 15)

Surgeon's knot (page 14)

Overhand knot (page 14)

7. End the thread by weaving through the core to meet the thread tail. Tie a surgeon's knot with the 2 threads and pass through a few size 8° beads, making half-hitch knots as you go.

8. Place crimp covers over all 4 crimp tubes.

Designer's Tip

As the size 8° beads begin to fill with thread, you may find it difficult to pass your needle through. In this case, pull the needle through with chain-nose pliers.

Dangling Bangle

A simple wire technique is used to create this elegant bracelet that changes with each movement of the wrist.

Designer: Mary Hettmansperger

Materials

39 to 42 light amber 4 mm crystal cube beads

75 to 81 dark charcoal brown 4 mm crystal bicone beads

39 to 42 black 3 mm crystal round beads

7 to 8 feet of 20-gauge half-hard sterling silver wire

39 to 42 sterling silver 2-inch head pins

Three-strand sterling silver slide-lock clasp

Tools

Wire cutters

Chain-nose pliers

Round-nose pliers

Techniques

Wrapped loop (page 21)

Note

The bracelet shown is approximately 8 inches long and fits an average wrist. Keep in mind that for a larger or smaller bracelet you may need to add or subtract links and bead dangles to get the proper fit.

Instructions

1. Cut 36 pieces of wire, each 2½ inches long. Set aside.

2. Use 1 head pin to string 1 cube bead, 1 bicone bead, and 1 round bead. Make a wrapped loop to secure the beads. Repeat to make 39 bead dangles. Set aside.

3. Begin a wrapped loop about one-third of the way down a piece of the cut wire. Before making the wrap, attach the loop to one of the bead dangles and to one of the holes on half of the clasp (figure 1). Complete the wrap.

fig. 1

4. String 1 bicone bead and make a wrapped loop on the other end of the wire to complete the link. The link should be about ¾ inch long.

5. As in step 3, begin a wrapped loop about one-third of the way down a piece of the cut wire. Before making the wrap, attach the loop to one of the bead dangles and to the open loop of the last link you made (figure 2).

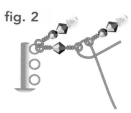

fig. 2

6. String 1 bicone bead and make a wrapped loop on the other end of the wire to complete the link.

7. Repeat steps 5 and 6 to make 9 or 10 more 2-part links.

8. Repeat steps 3 to 7 to add strands to the second and third clasp loops. Work to keep all 3 strands the same length. Test for fit, keeping in mind that you will be adding 1 more link length to connect to the other half of the clasp. Adjust the strands as needed.

9. Slide the clasp halves together.

10. Add the last link to the end of the first strand, but this time work the second wrapped loop of the link so it attaches to a bead dangle and the coordinating loop on the other half of the clasp. Repeat for each strand.

Confetti

A domed mesh clasp is a natural base for an explosion of crystals on this fanciful bracelet.

Designer: Nancy Zellers

Materials

Approximately 500 clear AB 3 mm crystal round beads

Approximately 50 size 8° pearl seed beads

Approximately 50 clear AB size 15° seed beads

Gold or silver ¾-inch mesh dome box clasp

15 to 20 grams rose-lined transparent 3 mm cube beads

Pink and white size D beading thread

Beeswax

Tools

Size 10 beading needle

Scissors

Flexible measuring tape

Techniques

Simple fringe (page 15)

Ladder stitch (page 17)

Herringbone stitch (page 16)

Instructions

Making the Clasp

1. Remove the mesh dome from the clasp.

2. Thread the needle with about 8 feet of white thread. Double the thread and wax it heavily. Secure the end of the thread to the center of the dome. Exit the thread on top of the dome.

3. **fig. 1**

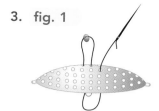

String 1 size 8° bead and pass down through the closest hole from where you last exited and up through an adjacent hole (figure 1). Repeat until you've covered the entire top of the dome with beads. Finish by beading a neat round of beads around the outside edge of the dome. Secure the thread and trim it close to the beads.

4. Thread the needle with about 5 feet of waxed white thread. Secure the thread so it exits from a seed bead at the top center of the dome.

5. **fig. 2**

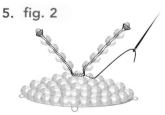

String 5 crystal beads and 1 size 15° bead. Skip the seed bead and pass back through the 5 crystal beads and the seed bead you last exited. String 4 crystal beads and 1 size 15° bead to create another fringe leg. Pass back through the same seed bead you last exited and through an adjacent seed bead (figure 2).

6. Repeat step 5, adding 2 fringe legs to every size 8° bead on the dome, working from the center of the dome outward. Secure the thread and trim close to the beads.

7. Fit the dome to the box to complete the clasp. Set aside.

Making the Bracelet

1. Measure your wrist and subtract the width of the clasp. Use this figure to determine the length of your bracelet. *Note:* The clasp will be heavy, so be sure to measure so the bracelet will be snug on your wrist.

2. Thread your needle with about 8 feet of pink thread. Double the thread and wax it heavily.

3. Use the cube beads to work a strip of ladder stitch 2 beads high and 6 beads wide. Exit up through a bead at the end of the ladder.

4. Work herringbone stitch off of the ladder. Begin by stringing 2 cube beads. Pass down through the adjacent bead stack on the ladder-stitched strip and up through the next. Repeat twice.

5. fig. 3  Step up to the next row by passing up through the second-to-last bead on the previous row, exiting between the previous row and the newly added row. Pass up through the last bead added in the newly added row (figure 3). This way of stepping up hides the thread within the bracelet as it strengthens the outside edges.

6. Continue to work herringbone stitch until the bracelet is ½ inch from the desired length. Don't cut the working thread.

7. Start a new thread at the beginning of the bracelet and securely sew on the embellished dome portion of the clasp. Check the bracelet for fit, keeping in mind that you have ½ inch more beading to do. Add or subtract herringbone rows if needed.

8. Use the working thread at the end of the bracelet to work a row of 2-bead-high ladder stitch across the end of the bracelet to match the beginning.

9. Securely sew the other half of the clasp to the end of the bracelet.

Chandeliers

Attach bead dangles to a wire component
to create an elegant chandelier effect.

Designer: Sandra Lupo

Materials

2 smoky topaz 12 x 8 mm crystal polygon beads

4 opaque aqua 8 mm crystal round beads

12 opaque aqua 4 mm crystal bicone beads

6 denim blue 6 mm crystal bicone beads

4 denim blue 3 mm crystal bicone beads

10 golden champagne 3 mm crystal bicone beads

18 gold-plated 4 mm daisy spacer beads

10 gold-filled 2-inch head pins

2 feet of 22-gauge round dead-soft gold-filled wire or gold-plated wire

Making Your Own Wireworking Jig

It's easy to make your own wireworking jig. All you need is a 6 x 6-inch wooden board and steel nails. First, use heavy-duty wire cutters to clip the heads off the nails, and then file them smooth with a needle file or emery paper. Hammer the headless nails into the board where the pattern indicates, making each headless nail a ⅝-inch tall peg. *Note:* To make a jig for the earrings in this project, place the pegs no more than 3 to 4 mm apart.

Instructions

Making the Wire Form

1. Make a homemade jig (see bottom, left) using peg pattern 3, or set up your commercially made jig with pattern 1, shown below.

fig. 1

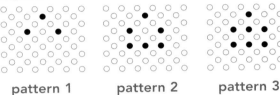

pattern 1   pattern 2   pattern 3

2. **fig. 2**

Cut a 9-inch piece of wire. Working with peg pattern 1, center the wire above the top peg and crisscross the wire underneath the peg. Curve the left-side wire under, around, and down the inside of the bottom left peg. Curve the right-side wire under, around, and down the inside of the bottom right peg (figure 2). Use the awl to gently tamp down the wire flat to the jig.

3. If you're using a commercially made jig, add pegs as needed to make pattern 2. Curve the left-side wire counterclockwise around the left peg on the bottom row of pegs. Use chain-nose pliers to tighten the wrap. Curve the right-side wire clockwise around the right peg on the bottom row of pegs. Tighten the wrap.

4. **fig. 3**

Curve the left-side wire clockwise around the center peg on the bottom row of pegs and tighten (figure 3). Curve the right-side wire counterclockwise around the same peg and tighten.

5. fig. 4

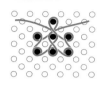

If you're using a commercially made jig, add a peg to make pattern 3. Move the left-side wire so it curves around the right side of the remaining center peg. Move the right-side wire so it curves around the left side of the remaining center peg (figure 4). Use the awl to tamp the wires into place.

6. Carefully lift the wire form from the pegs. If you've firmly tamped the wires down while forming the component, you should be able to lift it off in one piece. On the commercially made jig, you may remove a few pegs to remove the wire form.

7. Use nylon-jaw pliers to firmly grasp the form so only the top loop shows. Use chain-nose pliers to wrap one wire end tightly around the crossed wire beneath the loop. Repeat with the remaining wire end, wrapping in the opposite direction. Use the wire cutters to flush cut any excess wire close to the wrap. Use bent chain-nose pliers to gently squeeze the wrap, tucking the wire ends in.

8. Adjust the form as needed, flattening wraps with the nylon-jaw pliers and reshaping loops with the single jaw of round-nose pliers.

Repeat steps 1 to 8 to make a second wire form.

Creating Your Own Ear Wires

1. If you're using a commercially made jig, position 1 regular peg and 1 large peg one peg hole diagonally away from each other on the board.

2. Cut a 2½-inch piece of wire. Use round-nose pliers to form a loop at one end of the wire that will fit onto a regular peg.

Tools

Homemade wire jig (6 x 6-inch wooden board and steel nails that are 2 to 3 mm in diameter) or commercial wire jig with removable stainless steel pegs (including a ¼-inch-wide peg)

Hammer (optional)

Heavy-duty wire cutters (optional)

¼-inch-diameter dowel (optional)

Wire cutters

Awl or other thin steel rod

Chain-nose pliers

Nylon-jaw pliers

Bent chain-nose pliers

Round-nose pliers

Metal needle file or emery papers (medium and fine grits)

Techniques

Using a jig (11)

Simple loop (page 21)

Opening and closing loops (page 13)

Note

Each peg of the commercial jig should be 2 to 3 mm in diameter and spaced 3 to 4 mm apart.

3.

fig. 5

Place the loop on the regular peg on the board, and wrap the extra wire around the large peg (figure 5).

4. Cut any excess wire and file the wire end smooth. Use chain-nose pliers to make a slight bend in the wire away from the loop, about ¼ inch from the end.

5. Repeat steps 1 to 4 to make a second ear wire.

6. If you aren't using a commercially made jig, start with step 2. Hold the dowel in one hand and the wire loop in the other so it's perpendicular to the dowel. Wrap the straight end of the wire around the dowel to get a nice curve. Finish as in step 4. Repeat to make a second ear wire.

Making the Dangles

1. Use 1 head pin to string 1 polygon bead, 2 spacers, 2 opaque aqua 4 mm bicone beads, 1 spacer, 1 denim blue 6 mm bicone bead, and 1 golden champagne 3 mm bicone bead. Make a simple loop to secure the beads and set aside. Repeat to make a second long dangle.

2. Use 1 head pin to string 1 opaque aqua 8 mm round bead, 2 spacers, 1 denim blue 6 mm bicone bead, 1 opaque aqua 4 mm bicone bead, and 1 golden champagne 3 mm bicone bead. Make a simple loop to secure the beads and set aside. Repeat to make 3 more medium dangles.

3. Use 1 head pin to string 1 opaque aqua 4 mm bicone bead, 1 spacer, 1 denim blue 3 mm bicone bead, and 1 golden champagne 3 mm bicone bead. Make a simple loop to secure the beads and set aside. Repeat to make 3 more short dangles.

Assembling the Earrings

1. Attach the long dangle to the center bottom loop of one of the wire forms. Attach 1 medium dangle to each of the outside bottom loops of the form. Attach 1 short dangle to each of the outside bottom loops so they sit toward the outside of the form, away from the medium dangles.

2. Attach the ear wire to the top loop of the wire form.

3. Repeat steps 1 and 2 to assemble the second earring.

Gossamer

Assemble branches of flexible beading wire and crystals into an alluring necklace. Drop earrings complete the set.

Making the Necklace

Instructions

1. Cut 20 inches of wire. String 1 crimp tube, one 3 mm bead, 1 crimp tube, and the claw side of the clasp. Pass back through all the beads just strung, leaving a 1-inch tail. Snug the wire until the distance between the last crimp tube and the clasp is about 5 mm. Use flat-nose pliers to squeeze the tubes flat.

2. String 1 crimp tube and place it along the wire 4½ inches from the first crimp tube added in the previous step. Squeeze the tube flat.

3. String twelve 4 mm beads.

4. String a sequence of three 8 mm beads and one 4 mm bead eight times.

5. String eleven 4 mm beads and 1 crimp tube. Place the crimp tube along the wire 13½ inches from the first crimp tube added in step 1, and squeeze it flat.

6. String 1 crimp tube, one 3 mm bead, 1 crimp tube, and 1 small jump ring. Pass back through the beads just strung, leaving a 1-inch tail. Make sure the distance between the first crimp tube strung in this step and the one added in the last step is 4½ inches to match the first side of the necklace (figure 1).

   fig. 1

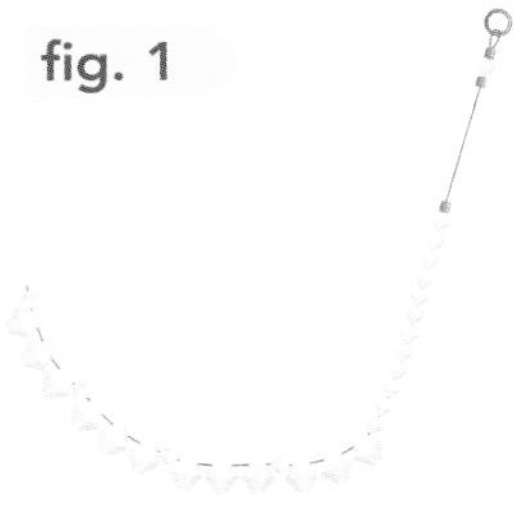

7. Open the large jump ring and attach it to the remaining 8 mm bead. Close the jump ring. Use a small jump ring to connect the bead dangle you just made to the small jump ring at the end of the necklace. Close the jump ring and set the strand aside.

8. Cut 13 pieces of wire 2½ inches each. Set aside.

9. Use one of the wire pieces to string 1 crimp tube and squeeze it flat near the end of the wire. String one 3 mm bead, 1 crimp tube, and one 4 mm bead. Pass the wire through the twelfth bead strung in step 3. String one 4 mm bead, 1 crimp tube, one 3 mm bead, and 1 crimp tube. Spread the crimp tubes and beads over the wire and, making sure the beads can slide freely along the wire, squeeze the tubes flat. This makes up 1 "branch" (figure 2).

   fig. 2

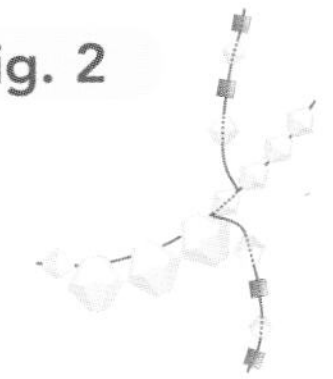

10. Repeat step 7 to add 12 more branches along the beads added in step 4. Space the branches more or less evenly apart, but give the branches a random look by sometimes using 4 instead of 3 crimp tubes, varying the space between the tubes, and passing through 4 mm beads in one place, and through 8 mm beads in another.

Designer: Katherine Song

Materials for Necklace

25 clear 8 mm top-drilled crystal bicone beads

62 clear 4 mm crystal bicone beads

28 clear 3 mm crystal bicone beads

2 gold-filled 4.5 mm jump rings

1 gold-filled 6 mm jump ring

58 to 68 gold-filled 1 x 1 mm crimp tubes

1 gold-filled 7 x 13 mm lobster clasp with soldered jump ring

52½ inches of .019 flexible beading wire

Tools

Wire cutters

Flat-nose pliers

Techniques

Stringing (page 14)

Making the Earrings

Instructions

1. Cut a 5-inch length of beading wire. String 1 crimp tube, one 4 mm bead, and 2 crimp tubes, leaving a 1-inch tail. Pass back through all to form a loop no larger than 4 mm long (figure 1).

fig. 1

2. Use the other end of the wire to string 1 crimp tube, one 3 mm bead, one 4 mm bead, three 8 mm beads, one 4 mm bead, one 3 mm bead, and 1 crimp tube.

3. Pass through the beads strung in step 1 (figure 2). Squeeze the crimp tubes added in step 1 flat and trim the wires close to the crimps.

fig. 2

4. Position the beads hanging on the large bottom loop so they are centered. Squeeze the crimp tubes added in step 2 flat so the beads stay in position.

5. Open a jump ring and attach the loop and 1 ear wire. Close the jump ring.

6. Repeat steps 1 through 5 to make the second earring.

Materials for Earrings

6 clear 8 mm top-drilled crystal bicone beads

6 clear 4 mm crystal bicone beads

2 clear 3 mm crystal bicone beads

2 gold-filled 4 mm jump rings

10 gold-filled 1 x 1 mm crimp tubes

2 gold-filled ear wires

10 inches of .019 flexible beading wire

Tools

Wire cutters

Flat-nose pliers

Techniques

Stringing (page 14)

Wispy

Cascading beaded chains are the perfect complement to bare shoulders.

Designer: Marlynn McNutt

Materials

56 aqua AB 4 mm crystal bicone beads

42 sterling silver 1-inch head pins

14 sterling silver 1-inch eye pins

12 sterling silver 2 mm oval jump rings

2 sterling silver 20 mm round earring components with 11 loops along the bottom, 1 loop in the center, and a hanging loop at the top

47 inches of sterling silver 1.5 mm flat cable chain

Tools

Wire cutters

Chain-nose pliers

Round-nose pliers

Techniques

Simple loop (page 21)

Opening and closing loops (page 13)

Instructions

Preparing the Materials

1. Cut 10 pieces of chain each 3¾ inches long. Cut 2 pieces each 2½ inches long. Cut 2 pieces each 2¼ inches long. Set all the chains aside.

2. Use 1 eye pin to string 1 bicone bead. Make a simple loop to secure the bead. Repeat to make 14 beaded links in all. Set aside.

3. Use 1 head pin to string 1 bicone bead. Make a simple loop to secure the bead. Repeat to make 42 beaded dangles in all. Set aside.

4. Attach 1 beaded link to each end of a 3¾-inch piece of chain. Repeat to make 6 long beaded chains in all. Set aside.

5. Attach 1 jump ring to each end of the remaining 3¾-inch pieces of chain. Set aside.

6. Attach 1 dangle to one end of a 2½-inch piece of chain. Attach 1 jump ring to the other end. Repeat to make 2 short beaded chains in all. Set aside.

7. Attach 20 dangles to a 2¼-inch piece of chain. Make sure they are spaced evenly down its length, and that one of the dangles hangs from the last chain link. Attach a jump ring to the opposite end of the chain (figure 1). Repeat to make 2 focal chains in all. Set aside.

fig. 1

Assembling the Earrings

1. Open a loop at one end of a 3¾-inch beaded chain. Attach the link's loop to the first loop of an earring component. Close the link's loop. Open the loop on the other end of the chain and attach it to the seventh loop of the earring component. Close the link's loop.

2. Open a jump ring at one end of an unbeaded 3¾-inch chain. Attach the jump ring to the second loop of the earring component and close the jump ring. Open the jump ring at the other end of the chain. Attach it to the eighth loop of the earring component and close the jump ring.

3. Repeat steps 1 and 2 to connect a 3¾-inch beaded chain to the third and ninth loops of the earring component; a 3¾-inch unbeaded chain to the fourth and tenth loops; and a 3¾-inch beaded chain to the fifth and eleventh loops. The chains should layer over one another and hang in swags (figure 2).

fig. 2

4. Open the jump ring of one of the 2½-inch beaded chains. Attach it to the center loop of the earring component and close the jump ring. Adjust the chain so it hangs in front of all the chain swags.

5. Open the jump ring of one of the focal chains. Attach it to the loop inside the earring component and close the jump ring. Adjust the chain so it hangs in front of all the chains.

6. Open a loop at one end of a beaded link. Attach it to the top loop of the earring component and close the loop. Attach the top loop of the link to an ear wire.

7. Repeat steps 1 through 6 to assemble the second earring.

Soft Mesh

Wire mesh ribbon is married with beads
and grounded by an embellished rosette.

Materials

1 blue/green AB 12 mm crystal shank button

4 emerald 8 mm crystal bicone beads

4 light brown 8 mm crystal bicone beads

8 brown 8 mm crystal bicone beads

10 brown 6 mm crystal bicone beads

2 emerald 6 mm crystal bicone beads

24 emerald 4 mm crystal bicone beads

2 opaque turquoise 4 mm crystal bicone beads

3 light brown 4 mm crystal bicone beads

3 brown 4 mm crystal bicone beads

2 translucent aqua 4 mm crystal bicone beads

2 copper 3 mm miracle beads

1 turquoise 3 mm miracle bead

5 grams of size 11° copper seed beads

50 assorted size (3 mm to 20 mm), shape (round, rondelle, bicone, coin), and type (crystal, ceramic, stone, plastic, wood) beads in turquoise, brown, brass, and copper, with holes large enough to fit over the collapsed mesh

2 antique brass 2-inch eye pins

2 antique brass 6 mm oval jump rings

1 brass 9 to 12 mm jump ring

2 antique brass 13 x 15 mm filigree end cones

1 antique brass 35 mm 3-layer filigree rosette with 15 mm open center

1 brass 20 mm lobster clasp with rhinestone insets and connecting ring

3 feet each of gold, turquoise, and brown fine wire mesh

Instructions

Embellishing the Pendant

1. Attach the 9 mm jump ring to the top of the rosette.

2. Cut a 3-foot length of line and thread it through the needle. Secure the end of the thread to the center back of the rosette, leaving a 6-inch tail. Pass up through the rosette, string the crystal button, and pass down through the rosette. Pass through the rosette and button once more to reinforce.

3. Pass up through the rosette to exit from the next layer of the filigree. String 1 seed bead, 1 emerald 4 mm crystal bicone bead, and 1 seed bead. Pass down through the rosette so the beads become nestled in the filigree (figure 1).

fig. 1

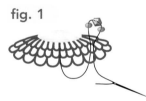

4. Repeat step 3 to evenly embellish the rosette with about 18 different 4 mm beads. When finished, pass down through the rosette where you began in step 2. Tie the working and tail thread together and trim.

Stringing the Necklace

1. Place a bead stopper or clip on one end of the flexible beading wire. String 1 emerald 4 mm bicone bead, 1 seed bead, 1 light brown 8 mm bicone bead, 1 seed bead, 1 emerald 4 mm bicone bead, 1 seed bead, 1 brown 6 mm bicone bead, 1 seed bead, 1 emerald 4 mm bicone bead, 1 seed bead, 1 brown 8 mm bicone bead, 1 seed bead, 1 emerald 8 mm bicone bead, 1 seed bead, 1 brown 8 mm bicone bead, 1 seed bead, 1 emerald 4 mm bicone bead, 1 seed bead, 1 brown

6 mm bicone bead, and 1 seed bead. Repeat the sequence three times. String 1 emerald 4 mm bicone bead, 1 seed bead, 1 light brown 8 mm bicone bead, 1 seed bead, 1 emerald 4 mm bicone bead, 1 seed bead, 1 brown 6 mm bicone bead, 1 seed bead, and 1 emerald 4 mm bicone bead. Clip the open wire end and set aside.

. Pass the ribbons, wire mesh pieces, and beaded strand lengths through the rosette's jump ring. Check that you have an equal amount of material exiting from each side of the jump ring, but don't cut anything yet. The mesh will reduce in length when you pull the width to achieve the ruffles.

. Twist the mesh into a tight cord with your index finger and thumb for about 1 inch. Use this length like a needle to string your first assorted bead onto the mesh. Slide the bead down toward the rosette. Repeat to place 5 assorted beads along the mesh on one side of the necklace, and then 5 assorted beads on the other side. Leave significant gaps between the beads. Note: The beads you string on the mesh should stay in position, but if you have a bead with a larger hole that slides, tie an overhand knot closely above and below it to keep it in place.

. When you are satisfied with the bead placement, use your fingers to pull and twist the mesh in opposite directions, giving a wavy, ruffled effect (figure 2). The necklace shown also uses another technique with the mesh. First tie a knot in the mesh, and push about 15 various-type 3 mm and 4 mm beads down the mesh tube. It may help to use the end of a paintbrush to get the beads into the tube. Tie a knot on top of the collection to create a little pocket of beads. Pull the mesh to give the pocket shape.

fig. 2

Materials (continued)
3 feet of 4 mm-wide satin ribbon

3 feet of 6 mm-wide organza ribbon

18 inches of .019 flexible beading wire

Braided beading line

Clear adhesive cement

Tools
Beading needles

Scissors

Round-nose pliers

Wire cutters

Bead stoppers or other clips

Techniques
Stringing (page 14)

Overhand knot (page 14)

Bead embroidery (see instructions)

Designer's Tip
When choosing the large-holed beads, bring a piece of 18-gauge wire with you to the bead shop. Then you can insert the wire through the bead to see if it will work for your project.

5. Repeat steps 3 and 4 for each of the wire mesh strands.

6. Trim along the end edge of the organza ribbon for 1 inch. Cut a diagonal across the ribbon. This shape will act as a needle to string beads. As you did with the mesh, string assorted beads and slide them down the ribbon, positioning them as desired. String 2 or 3 assorted beads on one side of the necklace, and 2 or 3 on the other. If any of your beads slide too easily, make a knot directly above and below it to keep it in place.

3. Hold all the strands together from one side of the necklace and loosely tie an overhand knot. Adjust the positioning of the knot, taking into account the desired length and the closure length. Repeat for the other side.

4. Pass 1 eye pin through the central hole of the knot at one end of the necklace. Tighten the knot around the eye pin. Check that the length is still accurate, make adjustments as necessary, and then pull the knot tight.

5. Open the eye pin wide and pull it over the knots, making sure that all of the strands are captured within the eye pin (figure 3). Tightly close the eye pin over the knots.

fig. 3

6. Cut the excess threads close to the eye pin. Dab the knots with enough glue so that the fabrics absorb the glue and the strands won't come undone.

7. String 1 end cone, 1 seed bead, 1 emerald 6 mm bicone bead, and 1 seed bead onto the eye pin. (The end cone should hide the knot.) Make a simple loop to secure the cones and beads.

8. Repeat steps 4 through 7 for the other side of the necklace.

9. Use jump rings to attach the clasp to the eye pins.

7. Cut the top of the satin ribbon at an angle. String 7 assorted beads for each side of the necklace. You may want to stick with large-hole beads only for this ribbon, because it frays easily. As before, tie knots above and below the beads that tend to slide.

Assembling the Necklace

1. Gather all the threads together and place the necklace around your neck to check for fit. Subtract the length of your clasp to reach a final measurement.

2. Adjust any beads that are not sitting in place on the strands.

Briolette Glam

Exploit the beauty of clear crystals,
showing off diamond-like briolettes.

Designer: Laura Shea

Materials

1 clear 18 mm crystal briolette pendant

2 clear 15 mm crystal briolette pendants

2 clear 13 mm crystal briolette pendants

179 clear 4 mm crystal bicone beads

2 clear 3 mm crystal bicone beads

1 sterling silver 13 mm lobster clasp with jump ring

1 sterling silver 5 x 8 mm double-loop connecting ring for lobster clasp

1 sterling silver 4 mm soldered jump ring

2 sterling silver 2 x 2 mm crimp beads

Monofilament, 6-pound test, or 6 to 8-pound test clear braided fishing line

20 inches of flexible beading wire

Tools

Scissors

Size 12 beading needles

Chain-nose pliers (optional)

Techniques

Angle stitching (see instructions)

Stringing (page 14)

Crimping (page 14)

Note

The materials listed are for a 16-inch necklace. Add or subtract size 4 mm bicone beads for a longer or shorter version.

Instructions

Note

If you are having trouble threading your needle with the monofilament, pinch the end with chain-nose pliers, and then trim it to a point with sharp scissors. You can also work without a needle. Be sure to pull the line securely through the beads, because these types of line can kink inside the holes of the crystal beads.

Stitching the Centerpiece

Follow figure 1 for thread path.

1. String 12 clear 4 mm bicone beads (beads 1 through 12). Pass through bead 1 to make a circle. Leave a 6-inch tail.

2. String 3 clear 4 mm bicone beads (beads 13, 14, and 15). Pass through beads 1 and 2.

3. String 4 clear 4 mm bicone beads (beads 16, 17, 18, and 19). Pass through beads 13, 2, and 3.

4. String 2 clear 4 mm bicone beads (beads 20 and 21). Pass through beads 16, 3, and 4.

5. String 1 clear 4 mm bicone bead, a 15 mm pendant, and clear 4 mm bicone beads (beads 22, 23 [15 mm pendant], 24, and 25). Pass through beads 20, 4, and 5.

6. String 2 clear 4 mm bicone beads (beads 26 and 27). Pass through beads 22, 5, and 6.

7. String 4 clear 4 mm bicone beads (beads 28, 29, 30, and 31). Pass through beads 26, 6, and 7.

8. String 1 clear 4 mm bicone bead, 1 clear 3 mm bicone bead, the 18 mm pendant, and 1 clear 3 mm bicone bead (beads 32, 33 [3 mm bead], 34 [18 mm pendant], 35 [3 mm bead]). Pass through beads 28, 7, and 8.

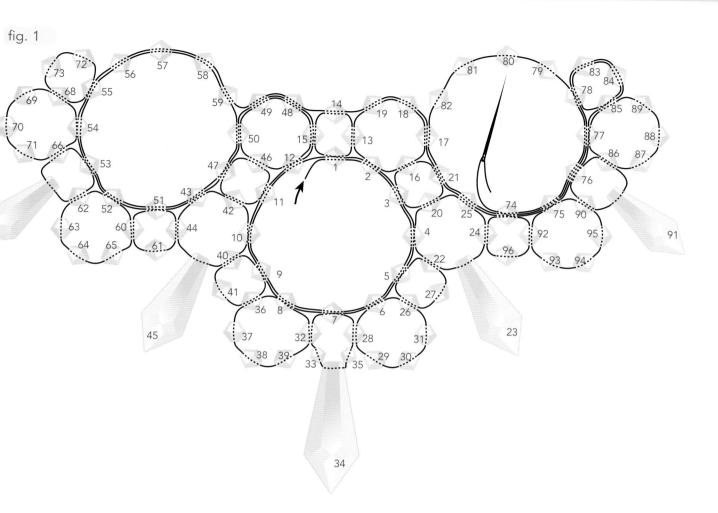

fig. 1

String 4 clear 4 mm bicone beads (beads 36, 37, 38, and 39). Pass through beads 32, 8, and 9.

0. String 2 clear 4 mm bicone beads (beads 40 and 41). Pass through beads 36, 9, and 10.

1. String 3 clear 4 mm bicone beads and a 15 mm pendant (beads 42, 43, 44, and 45 [15 mm pendant]). Pass through beads 40, 10, and 11.

2. String 2 clear 4 mm bicone beads (beads 46 and 47). Pass through beads 42, 11, 12, and 15.

3. String 3 clear 4 mm bicone beads (beads 48, 49, and 50). Pass through beads 46, 12, 15, 48, 49, 50, 47, and 43.

14. String 9 clear 4 mm bicone beads (beads 51 through 59). Pass through beads 50, 47, 43, and 51.

15. String 2 clear 4 mm bicone beads (beads 60 and 61). Pass through beads 44, 51, and 52.

16. String 4 clear 4 mm bicone beads (beads 62, 63, 64, and 65). Pass through beads 60, 52, and 53.

17. String 1 clear 4 mm bicone bead and a 13 mm pendant (beads 66 and 67 [13 mm pendant]). Pass through beads 62, 53, and 54.

18. String 4 clear 4 mm bicone beads (beads 68, 69, 70, and 71). Pass through beads 66, 54, and 55.

19. String 2 clear 4 mm bicone beads (beads 72 and 73). Pass through beads 68, 55, 56, 57, 58, 59, 49, 48, 14, 19, 18, 17, 21, and 25.

20. String 9 clear 4 mm bicone beads (beads 74 to 82). Pass through beads 17, 21, 25, 74, 75, 76, 77, and 78.

21. String 3 clear 4 mm bicone beads (beads 83, 84, and 85). Pass through beads 78, 83, 84, 85, and 77.

22. String 4 clear 4 mm bicone beads (beads 86, 87, 88, and 89). Pass through beads 85, 77, and 76.

23. String 1 clear 4 mm bicone bead and a 13 mm pendant (beads 90 and 91 [13 mm pendant]). Pass through beads 86, 76, and 75.

24. String 4 clear 4 mm bicone beads (beads 92, 93, 94, and 95). Pass through beads 90, 75, 74, and 24.

25. String 1 clear 4 mm bicone bead (bead 96). Pass through beads 92 and 74.

Finishing the Necklace

1. Weave through all the beads again to reinforce the beadwork. Secure the working and tail threads and trim close to the beads. Set the centerpiece aside.

2. Use the beading wire to string 1 crimp bead and the soldered ring. Pass back through the crimp bead, leaving a 1-inch tail.

3. String 45 clear 4 mm bicone beads. Slide the first few beads over both wires to cover the tail. Pass through beads 83, 79, 80, 81, 82, 18, 19, 14, 48, 49, 59, 58, 57, 56, and 72.

4. String 45 clear 4 mm bicone beads, 1 crimp tube, and the clasp connecting ring. Pass back through the crimp tube, snug all the beads tight, and crimp.

5. Attach the lobster clasp to the soldered ring.

Blue Persuasion

Water inspired this designer to create
a captivating bead and wire bracelet.

Materials

28 light blue 6 mm crystal round beads

44 sapphire 6 mm crystal round beads

44 clear 6 mm crystal round beads

5 feet of 20-gauge sterling silver wire (or 122 sterling silver 4 mm jump rings)

29 feet of 24-gauge sterling silver wire

1 sterling silver toggle clasp

Clear jeweler's cement for nonporous surfaces

Tools

Wire cutters

2.75 mm (size 2 US) knitting needle (for forming jump rings)

Flat-nose pliers

Chain-nose pliers

Round-nose pliers

Techniques

Coiling wire (page 21)

Opening and closing loops (page 13)

Wrapped loop (page 21)

Note

The materials are for a 6½-inch wrist. Add or subtract materials to make a larger or smaller bracelet.

Instructions

Making the Base Chain

1. Tightly coil the 20-gauge wire around the knitting needle down the needle's length. Slide the end of the coil off the end of the needle. Cut the coils one at a time to make jump rings. Make 122 jump rings in all.

2. Open 2 jump rings. Connect the 2 jump rings to 2 closed (stacked) jump rings. Repeat to make a chain 60 links long (figure 1).

fig. 1

3. Use 1 jump ring to attach half of the clasp to one end of the chain. Use 1 jump ring to attach the other half of the clasp to the opposite end of the chain.

Preparing the Dangles

1. Cut a 3-inch piece of 24-gauge wire. Make a wrapped loop at one end of the wire. String 1 clear bead and form a second wrapped loop, but don't cut the wire.

2. Hold the second loop with a pair of flat-nose pliers. Use chain-nose pliers to wrap the wire around and down the body of the bead until the wire meets the first loop. Loosely wrap the wire end around the first loop's base, and use chain-nose pliers to tighten the wrap (figure 2). Secure the wire end by adding a dab of glue where you've made the wrap. Repeat for each bead to make 116 dangles in all.

fig. 2

Assembling the Bracelet

1. Carefully open the second set of links at one end of the bracelet without dismantling the chain. Attach 1 clear dangle. Slide the links around so they open at the other side of the chain. Attach 1 clear dangle. You should end up with 1 dangle on each side of the chain (figure 3).

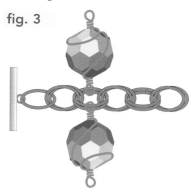

fig. 3

2. Repeat step 1 down the length of the chain, attaching 22 of the clear dangles, 22 of the sapphire dangles, and 28 of the deep blue dangles. As you work, move from one color to the next by softly blending the colors between the two links that separate them. For example, add 1 clear and 1 sapphire dangle on a set of links instead of 2 clear or 2 sapphire dangles.

3. Attach 22 sapphire dangles and 22 clear dangles until you come to the end of the chain.

Principessa

This glamorous necklace will make you feel as regal as an Italian princess.

Materials

8 clear 7 x 9 mm 2-hole crystal flatbacks

13 peridot 6 mm crystal round beads

155 clear AB 4 mm crystal bicone beads

Peridot size 8° hex beads

Bright silver size 11° true-cut seed beads

1 sterling silver 10 mm box clasp with semi-precious peridot inset

1 white 1 x 1-inch piece of stiff felt

White beading thread or beading line

Glue

Tools

Scissors

2 sharp beading needles

Techniques

Bead embroidery (see instructions)

Tubular peyote stitch (page 18)

Peyote stitch decreases (page 18)

Fringe (page 15)

Right-angle weave (pages 19 and 20)

Instructions

Making the Focal Piece

1. Dab a small amount of glue onto the back of 1 flatback and stick it to the center of the felt (don't let the glue ooze out the sides). Let dry.

2. Thread 1 needle with 3 feet of thread, and knot the end. Pass the needle up through the felt, right next to the flatback. String 3 seed beads, lay them on the felt along the edge of the flatback, and pass down through felt next to where the last bead lies. Pass up through the felt to pass through the second and third beads just added (figure 1). Repeat, working bead embroidery around the base of the flatback until you come to the first bead you added.

fig. 1

3. Pass through the first bead added in the base round. String 1 seed bead, skip a bead on the base round, and pass through the next bead. Continue, working a round of tubular peyote stitch. Step up to start the next round by passing through the last bead added in the base round and the first bead added in this round.

4. Work 3 more rounds of tubular peyote stitch, decreasing as necessary to form a tight bezel around the flatback's curved shape. Weave through the beads to exit from a bead on the base round. Set the needle aside.

5. Dab a small amount of glue on the back of 1 flatback. Stick it to the back of the felt, directly opposite the flatback already placed. Let dry. Trim the felt closely around the flatback, being very careful to avoid cutting any previous stitching.

6. Pick up the working thread. Work tubular peyote stitch on the other side of the base round, first working a straight round of tubular peyote stitch to cover the felt, and then making decreases as necessary to make a tight bezel around the second flatback (figure 2).

fig. 2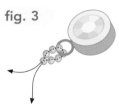

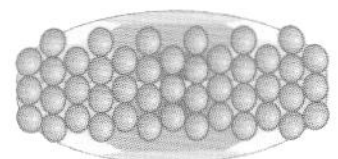

7. Weave through the beads to the base round. Exit from the top bead at one end of the bezel. String one 6 mm round bead and 1 seed bead. Pass back through the 6 mm round bead to make a fringe leg. Pass through the adjacent bead in the base round. Continue to add fringe around the base until you have 11 evenly spaced fringe legs in all. Secure the thread and trim close to the beads. Set the focal piece aside.

Creating the Straps

1. Thread 1 needle on each end of a 3-foot length of thread. String 6 seed beads and one half of the clasp. Pair the needles and string 1 seed bead (figure 3).

fig. 3

2. Use hex beads to work a strip of double-needle right-angle weave 16 units long.

3. String 1 bicone bead on the right needle and 1 bicone and 1 hex bead on the left needle. Cross the right needle through the hex bead. Repeat to make 6 more right-angle woven units.

4. Use the right needle to string 1 bicone bead, 1 hex bead, and 1 bicone bead. Pass through the right-hand hole on 1 flatback. Pass back through the beads just strung. Repeat with the left needle, passing through the left-hand hole of the flatback. Use both needles to reweave the units you made in this and the last step, reinforcing and securing the beadwork. Tie knots between beads if desired, and trim the threads close to the beads.

5. Thread 1 needle on each end of a 3-foot length of thread. Pass one needle through the left-hand hole on the flatback placed in the previous step. Pair the needles and string 1 bicone bead, 1 hex bead, and 1 bicone bead.

6. String 1 hex bead on the left needle and cross the right needle through it. String 1 bicone bead on the right needle and 1 bicone bead and 1 hex bead on the left needle. Cross the right needle through the hex bead. Repeat to make 8 more right-angle woven units.

7. Use the left needle to string 1 bicone bead, 1 hex bead, and 1 bicone bead. Pass through the left-hand hole on 1 more flatback, positioning the bead so it faces the same side up as the one placed in step 4 in this section. Pass back through the beads just strung. Repeat with the right needle, passing through the right-hand hole of the flatback. Use both needles to reweave the units, reinforcing and securing the beadwork.

As you weave up through the beadwork, stitch 1 hex bead over the flatback's left-hand hole (figure 4). Weave through several more beads, secure the thread, and trim close to the beads.

fig. 4

8. Repeat steps 5 through 7 for the right-hand hole on the first flatback. Set the strap aside.

9. Repeat steps 1 through 8 to create the strap for the other side of the necklace.

Connecting the Focal Piece

1. Lay out the necklace's components so the straps lie parallel, with the bottom 4 flatbacks in a row (flatbacks 1 through 4). Position the focal piece, best side up, below the 2 bottom middle flatbacks.

2. Thread 1 needle with a 3-foot length of thread. Leaving a 4-inch tail, pass through the right-hand hole of flatback 2. String 1 bicone bead, 1 hex bead,

1 bicone bead, 1 hex bead, and 1 bicone bead. Pass through the seed bead of the top fringe leg on the focal piece and back through the last bicone bead strung. String 1 hex bead, 1 bicone bead, 1 hex bead, and 1 bicone bead. Pass through the left-hand hole of flatback 3. Pass back through all the beads added in this step again to reinforce. When you return to flatback 2, string 1 hex bead and pass through the bicone on the other side of the flatback to cover the hole. Weave through the beads so you move down toward the first hole of flatback 2. Exit from the hole (figure 5).

fig. 5

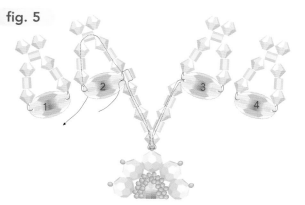

3. String 1 bicone bead, 1 hex bead, 1 bicone bead, 1 hex bead, and 1 bicone bead. Pass through the fringe leg just to the left of the top one and back through the beads just strung. When you return to flatback 2, cover the hole with a hex bead as before. Weave through beads so you exit from the third fringe leg to the left.

4. String 1 bicone bead and 1 hex bead. Repeat twice. Pass through the right-hand hole of flatback 1. Pass back through the beads just strung and into the fringe leg you last exited. Weave through beads on the focal piece to exit up through the fourth fringe leg to the left.

5. String 1 bicone bead and 1 hex bead. Repeat 4 times. String 1 bicone bead. Pass through the left-hand hole of flatback 1 and back through the beads just strung. Weave through the beads to place a hex bead over the holes of flatback 1.

6. Connect flatbacks 1 and 2 by stitching one 6 mm bead between the hex beads placed over the holes (figure 6). Secure the thread and trim.

fig. 6

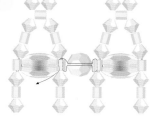

7. Repeat steps 3 through 6 to complete the other side of the necklace, working off flatbacks 3 and 4.

Mix & Match

These chic asymmetrical earrings can be combined in pairs for three different looks…what fun!

Making Earring 1

Designer: Katherine Song

Materials for Earring 1

1 clear AB 10 x 10 mm top-drilled (front to back) crystal heart bead

1 light topaz 6.5 mm crystal rhinestone charm with gold setting

4 smoky light topaz 4 mm crystal round beads

5 gold-filled 4 mm jump rings

1 gold-filled 1-inch head pin

1 gold-filled ear wire

5 inches of 28-gauge 14k gold-filled wire

Tools for All Earrings

Wire cutters

Round-nose pliers

Flat-nose pliers

Chain-nose pliers

Techniques for All Earrings

Simple loop (page 21)

Wrapped loop (page 21)

Opening and closing jump rings (page 13)

Twisted wire (see instructions)

Instructions

1. Make a simple loop at one end of the gold-filled wire.

2. String 1 round bead, the heart bead, and 1 round bead on the straight end of the wire. Working with the existing curve in the wire, pass the straight wire end through the looped end. Use flat-nose pliers to make a 90° bend in the wire so you form a 1-inch-diameter circle (figure 1).

   fig. 1

3. String 1 round bead onto the wire that sticks straight up from the simple loop. Make a wrapped loop to secure the bead. Set aside.

4. String 1 round bead onto the head pin and make a wrapped loop. Use 1 jump ring to attach the dangle to the wire loop, just to the left of the heart bead.

5. Link the remaining jump rings together. Attach one end of the chain to the charm. Attach the other end to the wire loop, just to the right of the heart bead.

6. Attach the earring to the ear wire.

Making Earring 2

Instructions

1. Make a simple loop at one end of the gold-filled wire.

2. String 1 burgundy 3 mm bead, 1 light brown 4 mm bead, and 1 light copper 5 mm bead onto the straight end of the wire.

3. Cut 10 inches of art wire and coil the end around the gold-filled wire a few times, about 1½ inches down from the loop. Position the new wire so you capture the beads strung in step 2 between the loop and the coil. Use chain-nose pliers to tighten the coil against the gold-filled wire.

4. String one 4 mm bead onto the new wire and hold it about ¼ inch from the coil. Bend the wire end down toward coil, hold on to the bead, and make a few twists (figure 2). Repeat to add the remaining 4 mm and 5 mm beads to the wire. Make a tight coil around the gold-filled wire to finish the art wire. Trim any excess art wire.

   fig. 2

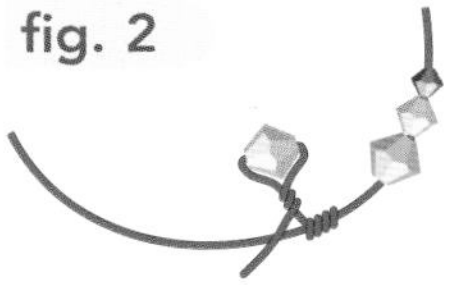

5. Use round-nose pliers to make a small loop at the end of the gold-filled wire. Grasp the loop with flat-nose pliers and turn the loop to make a flat spiral. Form a 1-inch-diameter loop with the wire.

6. Cut 1 inch of art wire and string it through the teardrop bead. Slide the bead so it is centered on the wire. Use your fingers to curve the wire ends along the top of the teardrop. Use chain-nose pliers to make the wire stick straight up from the top (figure 3). Use one of the wires to make a wrapped loop that uses the second wire as its stem (figure 4). Trim the excess wire or use round-nose pliers to coil it into a tiny spiral.

   fig. 3   fig. 4

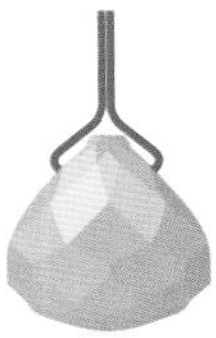

Materials for Earring 2

1 smoky light topaz 15 x 15 mm top-drilled crystal teardrop bead

1 light topaz 6.5 mm crystal rhinestone charm with gold setting

1 light peach 6 mm crystal bicone bead

5 light copper 5 mm crystal bicone beads

1 light brown AB 4 mm crystal bicone bead

2 smoky light topaz 4 mm crystal round beads

1 black 4 mm crystal round bead

3 deep burgundy AB 3 mm crystal bicone beads

3 gold-filled 4 mm jump rings

1 gold-filled ear wire

5½ inches of 28-gauge 14k gold-filled wire

13 inches of 28-gauge gold-colored art wire

7. Attach the dangle from step 6 to 1 jump ring. Attach the jump ring to the gold-filled wire between the loose crystal beads and the twisted ones.

8. Make a wrapped loop on one end of the remaining art wire. String 1 burgundy 3 mm bead. Make a wrapped loop at the other end that attaches to the charm. Attach 1 jump ring to the open wrapped loop.

9. Use 1 jump ring to attach the dangle made in step 7 and the wire loop to the ear wire.

Making Earring 3

Instructions

1. Cut 1 inch of art wire. Make a wrapped loop on one end. String 1 burgundy 3 mm bead and make a wrapped loop on the other end. Repeat to make another beaded link with 1 burgundy 3 mm bead, 1 beaded link with 1 black 4 mm bead, and 1 beaded link with the light brown 6 mm bead. Set aside.

2. Use the head pin to string 1 black 4 mm bead. Make a wrapped loop to secure the bead. Set the dangle aside.

3. Make 1 chain, using 2 jump rings to link the black dangle from step 2 to 1 burgundy beaded link to 1 black beaded link. End with 1 jump ring. Make another chain, using 2 jump rings to link the charm to 1 burgundy beaded link to 1 light brown beaded link. End with 1 jump ring. Set the chains aside.

4. Use 1 inch of art wire to create a dangle with the teardrop bead as you did for Earring 2 (see figures 3 and 4, page 126). Set aside.

5. Cut 15 inches of art wire. Use assorted beads to make a twisted wire and bead cluster (figure 5). End the cluster with a wrapped loop at the top and bottom.

fig. 5

6. Use 1 jump ring to attach the teardrop dangle to the bottom of the twisted wire and bead cluster, toward the back.

7. Attach the chain that includes the charm to the bottom of the cluster, just above the teardrop dangle and toward the back.

8. Use 1 jump ring to attach the remaining chain and the top of the cluster to the ear wire.

Materials for Earring 3

1 smoky light topaz 15 x 15 mm top-drilled crystal teardrop bead

1 light topaz 6.5 mm crystal rhinestone charm with gold setting

1 light brown 6 mm crystal bicone bead

3 light copper 5 mm crystal bicone beads

3 smoky light topaz 4 mm crystal round beads

3 black 4 mm crystal round beads

1 light brown AB 4 mm crystal bicone bead

1 smoky light topaz 4 mm crystal cube bead

4 deep burgundy 3 mm crystal round beads

1 gold-filled 4 mm round bead

1 gold-filled 1-inch head pin

8 gold-filled 4 mm jump rings

1 gold-filled ear wire

20 inches of 28-gauge gold art wire

Crystal & Fiber

Crystal dangles hang from
handmade silk scroll beads,
which are embellished with
crystal flat backs.

Designer: Kristal Wick

Materials

2 bronze 8 x 22 mm handmade silk scroll beads embellished with clear AB crystal flat backs

4 black 10 mm crystal bicone beads

4 smoky quartz 3 mm crystal bicone beads

Assortment of 48 beads, including 6 mm, 4 mm, and 3 mm crystal bicone and round beads in peridot, light smoky topaz, topaz, olive, light peach, copper, and smoky quartz

2 copper 5 mm barrel beads with raised design

2 sterling silver 3-inch eye pins

10 sterling silver 1½-inch head pins

2 sterling silver ear wires

2 pieces of sterling silver 3 x 6 mm figure-8 chain, each ¾ inch long

Tools

Wire cutters

Chain-nose pliers

Round-nose pliers

Technique

Simple loop (page 21)

Instructions

1. String an assortment of beads onto a head pin to equal ½ to 1 inch. Form a simple loop to secure the beads (figure 1). Repeat to make 5 beaded dangles.

fig. 1

2. Use chain-nose pliers to open the loop on 1 dangle and attach it to a link on one of the chain pieces. Close the loop. Repeat to attach 1 dangle to each chain link. Set the chain aside.

3. Open an eye pin and string 1 smoky quartz 3 mm bicone bead, 1 black 10 mm bicone bead, 1 silk scroll bead, 1 black 10 mm bicone bead, and 1 smoky quartz 3 mm bicone bead. Make a simple loop to secure the beads (figure 2).

fig. 2

4. Open one of the loops on the eye pin and attach an end link of the chain dangle.

5. Open the loop on one of the earring findings and attach the dangle. Close the loop.

6. Repeat steps 1 through 5 for the second earring.

I Thee Wed...

This necklace has a beautiful fringe on the back clasp, which is designed to grace the back of the bride's neck.

Materials

About 180 light blue 4 mm crystal round beads (A)

About 140 cream 5 mm crystal pearl round beads (B)

About 200 ivory 3 mm crystal pearl round beads (C)

Beading line, 10-pound test for the spiral and toggle and 6-pound test for the loop closure

½-inch piece of a round toothpick

Clear adhesive cement

Tools

Size 9, 10, or 11 metal crochet hook

Sizes 10 and 12 beading needles

Safety pins

Rubber bands

Thread burner or lighter

Techniques

Bead crochet (see instructions)

Instructions

Stringing the Beads

1. Lay out small amounts of each of the crystal beads. Thread a size 10 needle with the 10-pound line (thread) directly from the spool. You won't cut the thread until you are finished crocheting.

2. String the beads in a sequence of 1 A bead, 1 B bead, 1 A bead, and 1 C bead until you've strung 36 inches of beads. Make sure you complete the stringing with a full sequence.

3. Let out about 2 feet of thread and push all the beads down to the spool. Wrap half of the strung beads around the spool of thread it's attached to. Carefully secure the strung beads that have just been wound around the spool with a rubber band and wind the remainder of the strung beads (the ones you'll use first) on top of the other beads. This should leave you with the original 2 feet of thread, which will become the working thread. Make a slipknot at the end of the thread.

Crocheting the Beaded Tube

1. Put the hook into the slipknot and tighten the knot. Hold the slipknot on the hook with the index finger of your dominant hand. Slide a bead down toward the hook and wrap the working thread around your nondominant hand's index finger to prevent it from slipping. As you go on, you'll hold the work between your nondominant thumb and middle finger.

2. Make a yarn over. Capture the thread with the hook and pull it through the slipknot. You now have your first bead captured in your first stitch (figure 1). Repeat with the other 3 beads so you end up with 4 crocheted beads in a row. If you have kept the thread evenly tight, the chained row should curl a bit.

fig. 1

Bead Crochet Basics

1. Hold the crochet hook in your dominant hand between your thumb and third finger with the hook facing downward. This leaves your index finger on that hand free to manage the beads and thread.

2. Make a slipknot. Hold your thread with your nondominant hand between your thumb and index finger, leaving a 12-inch tail. The tail should sit at the back of your hand and the working thread in the front. Grasp the working thread with your dominant hand from the front and bring it behind the tail, creating a circle. Pull the working thread through this loop and insert the crochet hook into the loop (figure 2). Gently pull on the working thread until the loop closes around the hook. Use your dominant index finger to keep the loop in place on the hook.

fig. 2

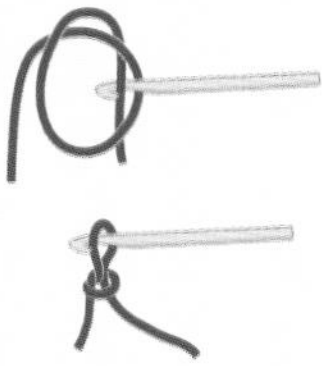

3. Loop the working thread over the index finger of your nondominant hand and hold it loosely across your palm with your last two fingers. Grasp the slip-knot with your thumb and middle finger so that you now hold the thread and hook with two hands.

4. You'll work slip stitch crochet to make the spiral tubes for this necklace. Slip stitch crochet consists of stitches that look like a series of Vs, or chain stitches (also called slip stitch-es). Hold the hook and thread in your hands, as above, and use your non-dominant index finger to draw the thread (yarn) over the crochet hook from back to front. This is called a yarn over (figure 3). Catch the thread with the hook by turning the hook slightly toward you so the yarn you brought over the hook from front to back slips into the groove of the hook. Draw the thread through the slipknot on the hook and up onto the working area of the hook (figure 4).

fig. 3

fig. 4

3. Position the tail so it sticks straight up from your work. Put the hook through the loop holding the first C bead you crocheted, the one closest to the tail. Maneuver the hook to the right of the bead so you have 2 loops on your hook. Yarn over, catch the thread with the hook, and pull it through the first and second loops (figure 5).

fig. 5

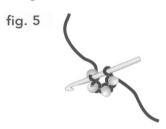

4. Work counterclockwise (if you are right-handed, clockwise if you are left-handed) as you crochet in rounds of 4 beads each. To make the next stitch, push the bead your hook is under away from you, so that the hook is on the left of the bead. You should now have 2 loops on your hook. Slide down the next bead (it should be the same type as the one your hook is under). Place the bead so that it sits over the 2 loops on your hook and over the bead your hook is now under.

5. Loop the thread over the hook and pull it through the loop holding the original bead as well as the loop on the hook. You are pulling the stitch through the 2 loops and will be left with the new single loop on the hook (figure 6).

fig. 6

6. Continue crocheting until you use up all the beads. Fasten off the work by making a chain stitch, and cut a sizable tail. Pull the end of the working thread through the loop. Set the bead-work aside.

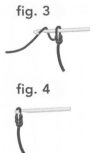

When you crochet a simple spiral tube like this one, you will always work a like bead into a like bead. Keep in mind that these first rounds are the most difficult and can be straightened out when you have finished by running your thread up and down through the like beads. Just like any other beadwork stitch, go slowly and be patient with yourself.

Also, if you find you've left out a bead when stringing, just make a stitch without the bead and you can sew the bead on later with a beading needle. If you strung one too many of one type of bead, put a safety pin in your working loop, put a needle or pin into the hole of the unwanted bead, shut your eyes, and use a flat-nosed pliers to carefully break the bead.

Adding Beads for Length

1. String more beads onto the spool thread. Form a slip stitch at the end, leaving a sizable tail, and attach it to the hook. Insert the hook from front to back into the last stitch and use it to join the new thread to the old one by making a slip stitch into the last stitch of the crocheted tube. Hang on to the two tails and continue to crochet with the new thread. Use a beading needle to weave the tails into the work in the direction from which it came.

2. When you've reached the desired length (the sample shown is 16 inches), make a slip stitch into each of the beads of the last round. Cut the thread from the spool, leaving a 12-inch tail, and pull the working thread through the last stitch.

Making the Toggle Bar

1. Thread a size 10 needle with 3 feet of 10-pound line. String 5 sequences of the A bead, B bead, A bead, C bead spiral pattern. Crochet a mini spiral tube that matches the larger spiral rope. Thread a beading needle on the thread and weave through all the beads again to reinforce. Dab the toothpick with glue and slip it inside the tube to help it remain stable. Let dry.

2. String 1 B bead and sew it to the end of the tube to hide the toothpick. Pass through the tube and repeat. Secure the thread, trim close to the bead, and burn the end. Set the toggle bar aside.

3. Start a new thread that exits from a C bead on the last 4-bead sequence at one end of the spiral. String 3 C beads and sew into a bead at the center of the toggle bar. String 3 C beads and pass through the C bead first exited from in this step (figure 7). Pass through all the beads again to reinforce. Secure the thread, trim close to the beads, and burn the thread end.

fig. 7

Making the Toggle Ring

1. Thread a size 12 needle with 3 feet of 6-pound thread. Secure it to the other end of the spiral rope, exiting from a C bead at the edge. String enough C beads that, when looped, can comfortably, but not too loosely, encircle the toggle bar attached to the other end of the spiral. Add a couple more C beads and pass back into the C bead you last exited.

2. Use C beads to work 2 rounds of peyote stitch off the loop of beads. Make a third round using B beads. Weave through to the first round of peyote stitch and work a round with C beads. Exit from the B bead on the last 4-bead sequence.

3. String 1 C bead, 1 B bead, and 1 C bead. Pass through the next C bead on the last peyote-stitched round. Repeat twice.

Adding the Fringe

1. String 1 A bead, 15 C beads, and 1 B bead. Skipping the last bead strung, pass back through the rest of the beads to make a simple fringe leg. Pass through the next bead on the peyote-stitched loop.

2. Repeat step 1, adding fringe legs with various bead combinations, around the outside of the loop. For a very full look, also make fringes off of the other peyote-stitched rounds.

Beaded Beads

Create crystal bead balls and string them to
make a stunning necklace and earrings.

Making the Bead Balls

Designer: Candie Cooper

Materials

32 olive 6 mm crystal bicone beads

60 deep red 5 mm crystal bicone beads

62 orange 4 mm glass bicone beads

50 red topaz 4 mm crystal bicone beads

52 brown 4 mm glass round pearl beads

72 topaz AB 4 mm crystal round beads

107 olive 4 mm crystal bicone beads

40 semiprecious red coral 4 mm round beads

22 smoky topaz 4 mm crystal bicone beads

22 copper AB 4 mm crystal round beads

62 deep red 4 mm crystal bicone beads

10 smoky topaz 4 mm crystal round beads

60 pearlescent brown 2 x 3 mm faceted beads

10 topaz 3 mm crystal bicone beads

52 deep red 3 mm crystal bicone beads

86 olive 3 mm crystal bicone beads

22 lime 3 mm crystal bicone beads

62 reddish brown 3 mm crystal bicone beads

62 smoky topaz AB 3 mm bicone Swarovski beads

2 sterling silver 2 x 2 mm crimp beads

2 sterling silver crimp bead covers (optional)

22 inches of .018 flexible beading wire

Clear nylon beading line, 0.35 mm diameter

Clear nylon beading line, 0.25 mm diameter

Clear adhesive cement

Instructions

Follow figure 1 for the thread path.

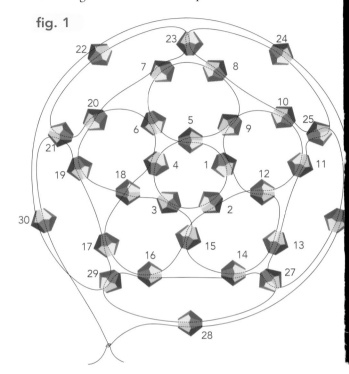

fig. 1

1. Cut 20 inches of 0.35 mm cord. Pair the ends so you have a "right thread" and a "left thread."

2. Begin by stringing 5 beads onto the left thread. Cross the right thread through the fifth bead strung, and snug the beads to make a star shape. These are beads 1 through 5.

3. String 4 beads (beads 6, 7, 8, and 9) onto the left thread. Cross the right thread through bead 9 and snug the beads to make a star shape.

4. Pass the right thread through bead 1. String 3 beads (beads 10, 11, and 12) onto the left thread. Cross the right end through bead 12 and snug the beads to make a star shape.

5. Pass the right thread through bead 2. String 3 beads (beads 13, 14, and 15) onto the left thread. Cross the right thread through bead 15 and snug the beads into shape.

6. Pass the right thread through bead 3. String 3 beads (beads 16, 17, and 18) onto the left thread. Cross the right thread through bead 18 and snug the beads.

7. Pass the right thread through beads 4 and 6. String 2 beads (beads 19 and 20) onto the left thread. Cross the right thread through bead 20 and snug the beads. The piece should now take on a domed shape.

8. Pass the right thread through bead 7. String 3 beads (beads 21, 22, and 23) onto the left thread. Cross the right thread through bead 23 and snug the beads.

9. Pass the right thread through beads 8 and 10. String 2 beads (beads 24 and 25) onto the left thread. Cross the right thread through bead 25.

10. Pass the right thread through beads 11 and 13. String 2 beads (beads 26 and 27) onto the left thread. Cross the right thread through bead 27.

11. Pass the right thread through beads 14 and 16. String 2 beads (beads 28 and 29) onto the left thread. Cross the right thread through bead 29.

12. String 1 bead (bead 30) onto the left thread. Pass through beads 22, 24, 26, and 28. Pass the right thread through beads 17, 19, 21, and 30. Tie the two tails together with a tight knot. Secure the knot with a dab of adhesive.

Tools

Scissors

Wire cutters

Crimping pliers

Chain-nose pliers

Techniques

Modified right-angle weave (see instructions)

Note

The instructions at left are for the basic pattern you'll use for all the woven balls. Follow the charts on the page that follows for information on the beads included in each type of ball and the order in which you'll use them.

Note

Use the 0.35 mm diameter nylon cord for the 6 mm and 4 mm beads and the 0.25 mm nylon cord for the smaller bead holes, such as the 3 mm crystals and red coral beads. Approximately 15 inches of cord is needed per bead ball for the 4 mm beads and 12 inches for the 3 mm beads.

The multicolored beads shown in the necklace are created by following the instructions for the beaded ball, but the beads are added in a certain order. The following lists the bead pattern specifics for each type of patterned ball.

Flower Cluster

Beads 1 through 5 are all one color and make up the flower cluster. Beads 6 through 30 are a different color.

Striped

To create the red coral/smoky topaz striped balls, follow this pattern.

Beads 1 through 5: red coral
Bead 6: smoky topaz
Beads 7 and 8: red coral
Bead 9: smoky topaz
Beads 10 and 11: red coral
Bead 12: smoky topaz
Beads 13 and 14: red coral
Bead 15: smoky topaz
Beads 16 and 17: red coral
Bead 18: smoky topaz
Beads 19 and 20: red coral
Beads 21 through 30: alternate between smoky topaz (bead 21) and red coral (bead 30).

Polka Dot

To create the red topaz/copper polka-dot balls, follow this pattern.

Beads 1 and 2: red topaz
Beads 3 through 8: alternate copper (bead 4) and red topaz (bead 8)
Beads 9 through 14: alternate red topaz (bead 9) and copper (bead 14)
Beads 15 and 16: red topaz
Bead 17: copper
Beads 18 and 19: red topaz
Beads 20 through 23: alternate copper (bead 20) and red topaz (bead 23)
Beads 24 and 25: red topaz
Bead 26: copper
Beads 27 through 29: red topaz
Bead 30: copper

Make 1 of the first and last beads listed and make the others in pairs.

# of Beads & Bead Color(s) per ball	Bead Shape	Bead Size	Pattern
30 olive	bicone	6 mm	solid
30 deep red	bicone	5 mm	solid
30 orange	bicone	4 mm	solid
5 red topaz/25 brown pearls	bicone/round	4 mm	orange flower cluster
5 topaz AB/25 olive	round/bicone	4 mm	topaz flower cluster
20 red coral/10 smoky topaz	round/bicone	4 mm	striped
19 red topaz/11 copper	bicone/round	4 mm	polka dot
30 deep red	bicone	4 mm	solid
5 smoky topaz/25 olive	round/bicone	4 mm	topaz flower cluster
30 topaz AB	round	4 mm	solid
30 faceted pearl beads	oblong	3 x 2 mm	solid
5 topaz/25 deep red	bicone	3 mm	topaz flower cluster
19 olive/11 lime	bicone	3 mm	polka dot
30 reddish brown	bicone	3 mm	solid
30 smoky topaz AB	bicone	3 mm	solid
30 olive (for clasp end)	bicone	3 mm	solid

Stringing the Necklace

Instructions

Stringing the Necklace

1. Once you've created all the balls, place them in stringing order according to the photo on page 135.

2. Use the beading wire to string 1 olive 4 mm bead, 1 crimp bead, and 1 olive 4 mm bead. Pass the wire through one of the star-shaped holes of the olive 3 mm ball, and pass back out through an adjoining hole. Pass back through the beads strung in this step (figure 2). Adjust the beads so the ball can move freely. Crimp the crimp bead. Trim any excess wire.

fig. 2

3. String the smoky topaz AB 3 mm ball onto the beading wire so the wire passes through two opposite star-shaped holes. String 1 smoky topaz AB 3 mm bead.

4. Continue stringing the balls. As you string, always follow each ball with a matching color and size crystal. There are two exceptions to this: the center olive 6 mm bead ball has olive 6 mm beads on each side of it. And there is no spacer bead between the pearlescent faceted bead ball and the topaz AB ball.

5. Adjust the stringing so that all of the spacer beads are nestled nicely between the woven balls.

6. String 1 olive 4 mm bead, 1 crimp bead, and 16 olive 3 mm beads. Pass back through the crimp bead and 4 mm bead to make a beaded loop. Snug the beads, crimp the crimp bead, and trim any excess wire.

7. Use chain-nose pliers to place crimp bead covers over the crimps, if desired.

Making the Earrings

Materials for Earrings

60 olive 4 mm crystal bicone beads

2 sterling silver long-legged ear wires

Clear nylon beading cord, 0.35 diameter

Tools

Scissors

Chain-nose pliers

Instructions

1. Make 2 olive 4 mm woven balls following the instructions on pages 136 and 137.

2. Slide 1 ball onto the ear wire through the small hole made by the three nylon cords intersecting between the beads and out through the hole on the opposite side.

3. Use chain-nose pliers to make a small hook at the end of the ear wire (figure 3) and slide the ball back down so one of the nylon cords catches on the hook.

fig. 3

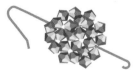

4. Repeat steps 2 and 3 to make the second earring.

Jean Campbell (Minneapolis, Minnesota)
Jean says she simply can't stop stitching, stringing, sewing, sleeping, and breathing beads! She is an author and editor whose specialty is beading, and she has written and edited dozens of books on the subject, including *The Beader's Companion* (with Judith Durant, published by Interweave Press), *Getting Started Stringing Beads* (Interweave Press), *Beaded Weddings* (Interweave Press), and *The Art of Beaded Beads* (Lark Books). She is the founding editor of *Beadwork* magazine.

Marie Lee Carter (New York, New York)
Marie began learning her craft in classes at the Fashion Institute of Technology and the 92nd Street Y in New York City. She focused on developing jewelry skills that allowed her to work without the chemicals and fumes of solder, pickle, and buffing compounds. In each of her designs, Marie aims to tell a story. (www.mariecarter.com)

Bonnie Clewans (Phoenix, Arizona)
Bonnie's interest in beading began when she was a young child. Her Ukrainian grandmother, who worked in the New York fashion industry, taught her beading, embroidery, and fine hand sewing. In 1993, she opened The Bead Gallery in Buffalo, New York, and began her professional career in beading. In addition to teaching in the shop, she began teaching at national beading conferences. She has written articles on beading for books and magazines, and she has been featured on DIY Network's *Jewelry Making*.

Candie Cooper (Shenzhen, China)
Candie graduated from Purdue University with degrees in fine arts and art education. Her passion lies in creating vibrantly colored jewelry from unique materials. Candie's jewelry has been exhibited throughout the United States, England, and Europe. She currently lives and works in China with her husband, Butch. (www.candiecooper.com)

Rachel Dow (Asheville, North Carolina)
Rachel received a BA in photography and an MA in art education from California State University, Northridge. She is a jewelry artist who specializes in fabricated silver, gold, and metal clay. She also enjoys dyeing and spinning fiber. (www.rmddesigns.com)

Anna Elizabeth Draeger (Waukesha, Wisconsin)
Anna began working with beads in 1990. A couple of years later, she walked through the door of a small bead shop in New York City and was awestruck when she saw a wall full of sparkling crystal beads. This experience convinced her to focus on making crystal jewelry. She currently works as an associate editor at *Bead&Button* magazine, teaches classes, and sells finished jewelry and instructions for many of her original designs. (www.beadivine.biz)

Peggy Gordon (New York, New York)
Peggy maintains a design studio in New York City, and she loves to teach beading, particularly bead crochet. Since 2000, her work has been judged into several competitions, including Swarovski's Create Your Style competition. Her work has been published in *Bead&Button* magazine as well as several beading books. Her work is represented by Snyderman-Works Galleries in Philadelphia. (sturpeg@aol.com)

Mary Hettmansperger (Peru, Indiana)
Mary is a fiber and jewelry artist who exhibits and teaches across the United States and abroad. Her work has been published in many magazines, including *Beadwork*, *Bead&Button*, and *Crafts Report*. Her work has been featured in several Lark books, including *500 Baskets* and *Fabulous Jewelry from Found Objects*. She is the author of Lark's *Fabulous Woven Jewelry*. (hetts@ctlnet.com)

Val Hirata (Honolulu, Hawaii)
Val was born and raised in Hawaii, and she feels fortunate to live in the warmth of a melting pot of cultures, surrounded by natural beauty. Her work reflects a blend of Japanese and European aesthetics. She confesses that she's a "Swarovski junkie" who is truly fascinated by the radiance of crystals. Currently, she designs beading projects for Creations By You, her family's business in Honolulu. (www.cbyweddings.com)

Tamara Honaman (Collegeville, Pennsylvania)
Tamara is a jewelry designer and media content manager for Fire Mountain Gems & Beads. She has appeared on the PBS series *Beads, Baubles & Jewels* and the DIY Network program *Jewelry Making*. She teaches jewelry making at national jewelry and bead shows and creates jewelry-making projects for international magazines. She is the founding editor of *Step by Step Beads* magazine. She has been creating jewelry for more than 12 years, working in a variety of media. (thonaman@msn.com)

Tina Koyama (Seattle, Washington)
Tina is a beadwork artist, instructor, and writer who has published designs in major beadwork magazines. Her work was included in the 2006 Bead International exhibition. She teaches regularly at Fusion Beads in Seattle and at national bead shows. (www.tinakoyama.com)

Elizabeth Larsen (Snohomish, Washington)
Elizabeth works as a biologist and creates jewelry to challenge her creative side. She is a self-taught beader who enjoys creating unique pieces of wearable art from sterling silver wire and semiprecious stones. Elizabeth's work is featured in Lark's *Contemporary Bead & Wire Jewelry*, and it has been published in *Bead&Button* magazine. (elarsen2003@yahoo.com)

Sandra Lupo (Lincroft, New Jersey)
Sandra has been making jewelry for 20 years. She teaches annually at Create Your Style with Swarovski, Beadfest, Jewelry Arts Expo, Wirefest, and Innovative Beads Expo. She has designed kits and finished jewelry for Touchstone Crystal and created projects for *Step by Step Beads* magazine. (sandra@sandstones.com)

Marlynn McNutt (Grants Pass, Oregon)
Marlynn first learned to bead from native Alaskan women and has been beading for 25 years. She moved from Alaska to work with Fire Mountain Gems & Beads as a designer and teacher. She has been involved with the PBS series *Beads, Baubles & Jewels* since its inception. Her work has been published in such magazines as *Simply Beads* and *Bead Unique*. (moosebreath76@charter.net)

Stacey Neilson (Dublin, Ireland)
Stacey has beading since she was 12 years old. She owns a successful bead and crystal shop in Dublin named Yellow Brick Road, where she teaches beading workshops and jewelry making. Her designs have been widely published in the United States and the United Kingdom. She is the author of the book *Jewellery-Making Basics*. ("Jewellery" is the proper spelling of "jewelry" in the United Kingdom.) (stacey@yellowbrickroad.ie)

Eni Oken (Los Angeles, California)
Originally from Brazil, Eni is a multidisciplinary artist whose work combines digital fantasy art, sculpture, jewelry, embroidery, and other mixed media. She holds a degree in architecture, and her art skills are self-taught. She notes that her professional career was focused on digital art for many years, but she chose to return to traditional media, such as jewelry design and painting, because of the tactile component that is missing in virtual art. Eni enjoys teaching and writing about jewelry making. (www.enioken.com)

Laura Shea (Denver, Colorado)
Laura specializes in beadwork based on and inspired by geometric patterns. She teaches beadwork in the United States and Europe. She incorporates crystal beads into practically every piece of her work. She loves their varied shapes, colors, and rainbow effects. She is currently working on her own book of geometrically based beadwork. For information about her shows and the classes she teaches, visit her website. (www.adancingrainbow.com)

Katherine Song (Toronto, Ontario)
Katherine moved from Xi'an, China, to Toronto in 2000. She holds a master's degree in fashion design. In 2005, she was awarded first place in the professional category of Swarovski's first annual Create Your Style design competition. Katherine described this masterful piece as "a piece of soul," a philosophical outlook that she conveys in all of her work. Wire twisting with Swarovski crystals is a hallmark of her unique design style. (www.katherinesong.com)

Christine Strube (St. Louis, Missouri)
Christine started working with beads five years ago when she began employment at Sorella Beads, a bead store and lampworking studio in St. Louis. Here she met and was inspired by many talented bead artists, including Stephanie Sersich, Dustin Tabor, Kate McKinnon, and Cindy Jenkins. Since then, she has begun teaching her own classes, selling jewelry at juried art shows, and publishing projects in magazines such as *Bead&Button, Step by Step Beads,* and *Beadwork*. (chstrube@earthlink.net)

Metric Conversion Chart

INCHES	METRIC (MM/CM)	INCHES	METRIC (MM/CM)	INCHES	METRIC (MM/CM)	INCHES	METRIC (MM/CM)
1/8	3 mm	1 1/2	3.8 cm	9	22.9 cm	16 1/2	41.9 cm
3/16	5 mm	2	5 cm	9 1/2	24.1 cm	17	43.2 cm
1/4	6 mm	2 1/2	6.4 cm	10	25.4 cm	17 1/2	44.5 cm
5/16	8 mm	3	7.6 cm	10 1/2	26.7 cm	18	45.7 cm
3/8	9.5 mm	3 1/2	8.9 cm	11	27.9 cm	18 1/2	47 cm
7/16	1.1 cm	4	10.2 cm	11 1/2	29.2 cm	19	48.3 cm
1/2	1.3 cm	4 1/2	11.4 cm	12	30.5 cm	19 1/2	49.5 cm
9/16	1.4 cm	5	12.7 cm	12 1/2	31.8 cm	20	50.8 cm
5/8	1.6 cm	5 1/2	14 cm	13	33 cm	20 1/2	52 cm
11/16	1.7 cm	6	15.2 cm	13 1/2	34.3 cm	21	53.3 cm
3/4	1.9 cm	6 1/2	16.5 cm	14	35.6 cm	21 1/2	54.6 cm
13/16	2.1 cm	7	17.8 cm	14 1/2	36.8 cm	22	55 cm
7/8	2.2 cm	7 1/2	19 cm	15	38.1 cm	22 1/2	57.2 cm
15/16	2.4 cm	8	20.3 cm	15 1/2	39.4 cm	23	58.4 cm
1	2.5 cm	8 1/2	21.6 cm	16	40.6 cm	24	61 cm

Terry Stumpf (Columbia, Illinois)
Terry is the assistant manager at Lady Bug Beads in St Louis, Missouri. She graduated with a degree in art and design from Columbia College in Chicago, where she studied jewelry making and metalsmithing. (terrylynn21@hotmail.com)

Karli Sullivan (Surprise, Arizona)
Karli has been beading for several years. She and her mother own Confetti: The Bead Place in Surprise, Arizona. She holds a degree in elementary education and now teaches all aspects of beading. Her work has been published in *Bead Style* and *Beadwork* magazines. (madredolce@yahoo.com)

Betcey Ventrella (Rollinsville, Colorado)
Betcey has been involved with beads for almost 25 years. In 1987, when she says it became apparent she wanted to own every bead she possibly could, she decided to open a bead store in Woodstock, New York. The store was open for 11 years before she and her family moved to Colorado. She is an avid collector of vintage Swarovski beads and stones. (www.beyondbeadery.com)

Wendy Witchner (Anchorage, Alaska)
Wendy is constantly on the road traveling to shows, while living and working out of a motor home. She shows three lines of jewelry that include wire and mixed metal. The limitations of life on the road, with minimal tools at hand, has led her to use primarily cold connections and wire in her work. She incorporates an eclectic mix of materials into her designs, including crystals, bali silver, pearls, and antique buttons from the late 1800s.

Nancy Zellers (Aurora, Colorado)
Nancy's work has been featured in many magazines, including *Beadwork* and *Bead&Button*. Her projects are included in several beading books published by Interweave Press. Her work is featured in *500 Beaded Objects* and *The Art of Beaded Beads*, published by Lark Books. Her sculptural pieces have been shown in numerous art exhibitions, and she enjoys teaching beadwork classes at conferences. (www.nzbeads.com)

Index